Edge Assisted Mobile Visual SLAM

Jingao Xu • Zheng Yang • Yunhao Liu • Hao Cao

Edge Assisted Mobile Visual SLAM

Springer

Jingao Xu
Tsinghua University
Beijing, China

Zheng Yang
Tsinghua University
Beijing, China

Yunhao Liu
Tsinghua University
Beijing, China

Hao Cao
Tsinghua University
Beijing, China

ISBN 978-981-97-3572-3 ISBN 978-981-97-3573-0 (eBook)
https://doi.org/10.1007/978-981-97-3573-0

This Springer imprint is published by the registered company Springer Nature Singapore Pte Ltd.
The registered company address is: 152 Beach Road, #21-01/04 Gateway East, Singapore 189721, Singapore

Life is like a box of chocolates, you never know what you're going to get.

To all of the students and alumni from our MobiSense group at Tsinghua University.

Preface

Visual Simultaneous Localization and Mapping (SLAM) has risen to prominence in recent years as a foundational technology for autonomous systems, augmented reality (AR) applications, and robotics. It serves as the eyes of machines, allowing them to navigate and understand the world with a level of precision and autonomy that mirrors human interaction. The importance of visual SLAM cannot be overstated, yet its full potential is often curtailed by the significant computational demands it places on mobile devices. As the expectations for real-time processing in autonomous vehicles, AR, and mobile robotics continue to escalate, the computational overhead of visual SLAM remains a critical bottleneck.

In this climate, the ascent of edge computing offers a new horizon for visual SLAM applications. By leveraging the processing power available at the edge of the network, closer to the data source, edge computing can alleviate the computational load from mobile devices, facilitating real-time operation and enhanced responsiveness. This monograph seeks to bridge the gap between the theoretical underpinnings of visual SLAM and the practical advancements enabled by edge computing. It is a timely response to the research community's call for innovative solutions that can harness the strengths of both domains to overcome the limitations of current systems.

The contents of this book are carefully curated to provide a comprehensive survey of the current landscape of edge-assisted visual SLAM. It begins with an overview of the principles of visual SLAM and progresses to explore how edge computing can augment these systems for improved efficiency and accuracy. The monograph highlights methods such as distributed computing paradigms, collaborative processing techniques, and adaptive algorithms that optimize data flow between mobile devices and edge servers. Special attention is given to the topics that are expected to pique the interest of readers, including the seamless integration of edge computing in SLAM for AR, autonomous navigation, the development of robust and scalable systems capable of operating in diverse environments, and the exploration of novel applications in mobile robotics.

Book Organization

The book guides the reader through the conceptual underpinnings to the latest innovations and applications of edge-assisted visual SLAM. It is structured to provide a logical flow from fundamental concepts to advanced applications, culminating in a forward-looking perspective on future research directions.

Part I lays the foundation with two chapters. Chapter 1 not only introduces the core principles of visual SLAM but also confronts the challenges inherent in executing such sophisticated algorithms in real-time on mobile devices. Chapter 2 discusses how edge computing could serve as a catalyst for visual tasks, facilitating real-time processing and reducing on-device computational burdens.

Part II focuses on the architectural advancements that make edge-assisted visual SLAM viable and effective. Chapter 3 presents the overarching system architecture, illustrating how edge computing and visual SLAM are integrated. Chapter 4 explores the challenges of scaling these systems for multi-agent environments and the solutions that current research provides.

Part III showcases practical applications of edge-assisted visual SLAM in various domains. Chapter 5 illustrates how these technologies are revolutionizing autonomous navigation. Chapter 6 delves into how to construct large-scale indoor navigation path from crowdsourced visual traces. Chapter 7 discusses the critical role of edge-assisted visual SLAM in understanding and interacting with complex environments. Chapter 8 demonstrates the application of visual SLAM in enhancing multi-user AR experiences.

Part IV offers a synthesis of the discussed topics and presents open questions that propel the field forward.

The organization of the book tries to ensure a progressive build-up of knowledge, with each chapter setting the stage for the next, allowing readers to develop a comprehensive understanding of the field. The goal of each chapter is not only to inform but also to inspire further innovation in edge-assisted visual SLAM.

Anticipated Audience

Readers will emerge with actionable insights into the design, implementation, and optimization of edge-assisted visual SLAM systems. They will learn not only of the technological advancements but also of the practical considerations that accompany the deployment of such systems. Readers will benefit from the detailed examination of state-of-the-art methods, results, and topics that have immediate relevance to current research and applications.

The monograph is intended for professionals, researchers, and students with a foundational understanding of SLAM, robotics, and computing systems. While prior knowledge of edge computing and its applications in mobile systems will be beneficial, the book is designed to be accessible to those with a basic understanding of these areas, providing a stepping stone toward more advanced study and research in this rapidly advancing field.

Beijing, China
March, 2024

Jingao Xu
Zheng Yang
Yunhao Liu
Hao Liu

Acknowledgments

Throughout nearly 5 years of dedicated research on edge-assisted visual SLAM systems, we have gained immense knowledge and support from our friends and colleagues. We seize this moment to express our profound gratitude to them.

First and foremost, we would like to express our deep gratitude to Prof. Chenshu Wu from the Hong Kong University, Prof. Longfei Shangguan from the University of Pittsburgh, and Prof. Kun Qian from the University of Virginia. Most of the works mentioned in this book were accomplished in close collaboration with them. They have been steadfast partners in our exploration of the emerging field of edge-assisted Visual SLAM.

Many former and present graduate students have contributed greatly to the book. Some chapters are based entirely on their writing. We particularly thank Erqun Dong, Liang Dong, Kehong Huang, Xinjun Cai, Jianzhe Liang, Danyang Li, Xiaowu He, Chen Qian, Zeyu Wang, Jialin Zhang, and Xiang Huang from the MobiSense Group. Without their hard work and strong support, this book would not have been possible.

We are grateful to many colleagues and peer researchers who shared with us exciting information. Especially, we thank Fan Dang, Xu Wang, Chaofan Yang, Changxu Wang, Tianmeng Hang, Yue Zheng, Tongtong Liu, Jing Chu, Kai Zhang, Yi Zhao, Hengjie Chen, Yumeng Lu, Min Sun, Yi Zhang, Guidong Zhang, Junjie Yin, Guoxuan Chi, Yue Wu, Qian Zhang, Kuntian Deng, and Hongbo Lan.

We thank all of those who have played in part in the preparation of this text. They include Prof. Mahadev Satyanarayanan, Prof. Ramesh Govindan, Prof. Karthik Dantu, Dr. Qiang Ma, Dr. Xin Miao, Mr. Zhuo Liu, Dr. Kebin Liu, and Prof. Xinlei Chen. The editors of this book, Wei Zhu and Sudha Ramachandran, of Springer provided many aspects of editorial support.

Finally, Jingao Xu thanks his wife Molly, and their parents Hui Xu, Chao Jia, Guangli Guo, and Xiaoying Zhang for their unconditional support. Hao Cao extends his deepest gratitude to Daisy Fan, whose love, care, and unwavering support were indispensable during his Ph.D. studies. We affectionately dedicate this book to them.

Contents

Part IV Conclusion

About the Authors

Jingao Xu received his Ph.D. and B.E. degree both from the School of Software, Tsinghua University, in 2017 and 2022. He is currently a postdoctoral researcher in the Computer Science Department, Carnegie Mellon University, after working as a postdoctoral researcher in the School of Software, Tsinghua University. His research interests include Internet of Things, mobile and edge computing, and visual SLAM. He is a member of the IEEE.

Zheng Yang is an associate professor in the School of Software and TNList, Tsinghua University. He received his B.E. degree in the Department of Computer Science from Tsinghua University, and his Ph.D. degree in the Department of Computer Science and Engineering of Hong Kong University of Science and Technology. Zheng received China National Natural Science Award (2011). His research interests include Internet of Things, computer networking, mobile computing, smart city, blockchain, etc. He is an author and co-author of 3 books and over 100 research papers in premier journals and conferences. He received 4 best paper (candidates) awards and has over 13,000 citations with H-index 58.

Yunhao Liu received his B.E. degree from the Department of Automation, Tsinghua University in 1995. He received his M.S. and Ph.D. degrees in computer science and engineering from Michigan State University in 2003 and 2004, respectively. He is a professor in the Department of Automation and the dean of the Global Innovation Exchange, Tsinghua University. He is a fellow of CCF, ACM, and IEEE. He is the Editor-in-Chief of *ACM Transactions on Sensor Networks and Communications* of the CCF. His research interests include Internet of Things, wireless sensor networks, indoor localization, the Industrial Internet, and cloud computing. He is an author and co-author of 5 books and over 200 research papers in premier journals and conferences. He has over 39,700 citations with H-index 90.

Hao Cao received his B.E. degree from the College of Intelligence and Computing, Tianjin University, in 2019. He is currently working toward the Ph.D. degree at the School of Software, Tsinghua University. His research interests include Internet of Things and mobile computing.

Acronyms

API	Application Interface
AR	Augmented Reality
ATE	Absolute Trajectory Error
BA	Bundle Adjustment
CDF	Cumulative Distribution Function
CNN	Convolutional Neural Network
DNN	Deep Neural Network
DoF	Degree of Freedom
EKF	Enhanced Kalman Filter
FPS	Frame per Second
GPS	Global Positioning System
HMM	Hidden Markov Model
IMU	Inertial Measurement Unit
IoT	Internet of Things
KCF	Kernel Correction Filter
LiDAR	Light Detection and Ranging Radar
LoS	Line of Sight
MAP	Maximum a Posteriori
MR	Mixed Reality
NeRF	Neural Radiance Field
PF	Particle Filter
RoI	Region of Interest
RRE	Relative Rotation Error
RTE	Relative Translation Error
SLAM	Simultaneous Localization and Mapping
VIO	Visual-Inertial Odometry
VO	Visual Odometry
VR	Virtual Reality

Part I
The Background

Part I lays the foundation for our exploration of edge-assisted Visual SLAM, providing the reader with essential background information and conceptual frameworks that are central to understanding and leveraging these powerful technologies in mobile environments.

Comprising two chapters, Part I begins by unpacking the essentials of Visual SLAM in Chap. 1. Here, we delve into its overview, examining each key component of Visual SLAM systems. We also pose some crucial challenges about implementing Visual SLAM on mobile platforms, where constraints such as computational power and real-time processing demand innovative solutions.

In Chap. 2, we shift our focus to Edge Computing, discussing how it integrates with mobile visual systems to enhance and facilitate real-time visual tasks. This chapter further elaborates on how Edge Computing can specifically enhance visual tasks, opening up new possibilities for advanced visual processing on mobile devices.

Chapter 1
Understanding Visual SLAM

Abstract From the inception of computer vision, there has always been a dream that one day, intelligent hardware devices could, like humans, observe the world through their *eyes*, perceive objects around them, comprehend spatial environments, and explore the unknown. This enchanting and romantic vision has captivated countless researchers, driving them to tirelessly pursue it night and day. Although achieving this has proven challenging, in today's era of increasing automation, many scenarios necessitate devices having the capability for spatial perception and positioning: indoor robotic vacuums need to model their working environment, autonomous vehicles on the roads must understand and locate themselves within those roads, drones in the sky require real-time tracking of their position and orientation, and virtual and augmented reality devices need to capture their viewpoint and comprehend real-world scenes. Without this capability, all these innovative entities could not manifest in our reality, which would be a significant loss.

To enable computers to parse spatial environments and ascertain their position within them, SLAM technology has seen significant advances over the past 30 years. SLAM stands for **S**imultaneous **L**ocalization **A**nd **M**apping, which involves an entity equipped with specific sensors, creating a model of its environment while in motion, without prior information about the environment, and simultaneously estimating its own movement. If the primary sensor is a camera, the technology is referred to as **Visual SLAM**. This chapter will primarily focus on Visual SLAM, the most widely applied form of SLAM in the industrial sector, to introduce this spatial intelligence computing technology.

1.1 Overview of Visual SLAM

The classical visual SLAM framework, as shown in Fig. 1.1, encompasses the following steps in its process:

- **Input: Sensor Data**: In visual SLAM, this primarily involves the reading and pre-processing of camera image data. Additionally, auxiliary inputs from the

J. Xu et al., *Edge Assisted Mobile Visual SLAM*,
https://doi.org/10.1007/978-981-97-3573-0_1

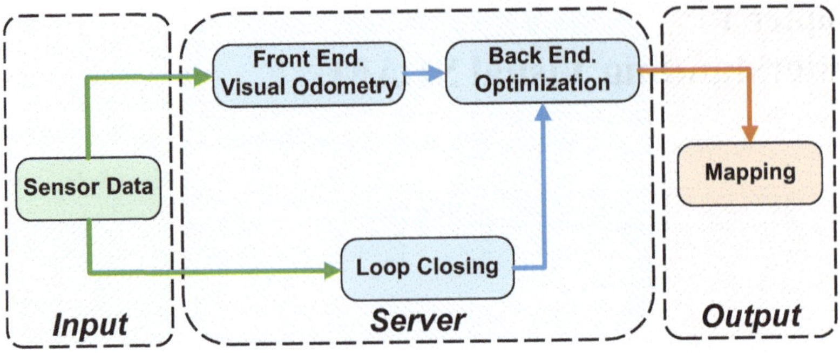

Fig. 1.1 Modules in visual SLAM

robot's onboard sensors (such as inertial sensors) can provide perceptions of its own state, including motion speed, acceleration, and angular velocity.

- **Front-end: Visual Odometry (VO)**: The task of visual odometry is to estimate the motion between the camera across two adjacent frames based on their differences, thereby calculating the camera's pose. VO is also known as the front-end of the SLAM system.
- **Back-end: (Non-linear) Optimization**: The back-end receives camera poses measured at different times by the visual odometry and information from loop closure detection, performing a joint optimization to obtain a globally consistent trajectory and map. Since it follows the VO, it is also referred to as the back-end.
- **Global: Loop Closure Detection**: Loop closure detection determines whether the camera has returned to a previously visited location. If a loop is detected, the information is forwarded to the back-end for processing to eliminate the cumulative errors in localization and mapping.
- **Output: Mapping**: Based on the estimated trajectory, a map corresponding to the requirements of the task is constructed.

In what follows, we will introduce each of these components in detail.

1.2 Key Components in Visual SLAM Systems

1.2.1 Input: Sensor Data

In visual SLAM, cameras capture the surrounding environment at a certain rate. Standard cameras can collect images at a speed of 30 frames per second, while high-speed cameras are even faster. In daily life, the commercial cameras we often encounter mainly fall into three categories: monocular (Monocular), stereo (Stereo), and depth cameras (RGB-D). Intuitively, a monocular camera has one lens, and stereo has two. The principle of RGB-D is more complex—in addition to capturing color pictures, it can also read the distance of each pixel to the camera. It usually carries multiple cameras, and its working principle differs from that of ordinary cameras.

Monocular Cameras The use of a single camera for SLAM is referred to as Monocular SLAM. This sensor configuration is simple and low-cost, making Monocular SLAM a subject of widespread interest among researchers. Essentially, a photograph is a projection of the scene at the moment of capture onto the camera's imaging plane. It reflects the three-dimensional world in two dimensions. Clearly, this process loses one dimension of the scene: the so-called depth (or distance). In monocular cameras, it is impossible to calculate the distance of objects in the scene from us through a single picture. Therefore, estimating the motion (Motion) of the body and the structure (Structure) of the scene both lack real scale (Scale). For example, if the camera's movement and the scene's size are both doubled, the monocular perspective remains unchanged. This deficiency is termed *scale ambiguity*, which poses significant challenges for the application of Monocular SLAM. The fundamental reason is the inability to determine depth from a single image. Therefore, to obtain this depth, people have begun to explore using stereo and depth cameras.

Stereo Cameras and Depth Cameras The purpose of stereo and depth cameras is to measure the distance to objects by some means, overcoming the drawback of monocular cameras not knowing the distance. If the distance is known, the three-dimensional structure of the scene can be recovered from a single image, thus eliminating scale ambiguity. Although both aim to measure distance, stereo cameras and depth cameras measure depth on different principles. A stereo camera consists of two monocular cameras, but the distance between these two cameras (known as the Baseline) is known. We estimate the spatial position of each pixel through this baseline—very similar to human vision. Humans can judge the distance of objects through the difference between images seen by the left and right eyes. The depth range measured by stereo cameras is related to the baseline. The greater the baseline distance, the greater the depth range that can be measured. The disadvantages of stereo or multi-lens cameras include complex configuration and calibration, depth range and accuracy limitations by the stereo's baseline and resolution, and significant computational resource consumption for disparity calculation, requiring acceleration with GPU and FPGA devices for real-time output of distance information for the entire image. Thus, computational load remains one of the main issues for stereo cameras under current conditions.

Depth Cameras (also known as RGB-D Cameras) emerged around 2010. They emit infrared light toward objects and receive the returned infrared light, using Time-of-Flight (ToF) to measure the distance from the objects to the camera. Unlike stereo cameras that rely on software calculations, this method involves physical measurements, thus significantly reducing computation. Commonly used RGB-D cameras include Kinect/Kinect V2, Xtion live pro, Realsense, etc., and the distance detection signals they emit are not limited to infrared light, even including lasers, millimeter waves, etc. However, most RGB-D cameras still have many problems, such as narrow measurement range, high noise, small field of view, susceptibility to sunlight interference, and inability to measure translucent materials. In SLAM, they are mainly used indoors, while outdoor applications are more challenging.

Onboard Sensors In addition to the cameras mentioned above as external sensors for environmental perception to infer one's position, we can also use the robot's own onboard sensors to estimate its motion internally, such as wheel encoders, Inertial Measurement Units (IMU), etc. These measure physical quantities that indirectly reflect the object's motion to obtain direct position data. For example, wheel encoders detect the angle of wheel rotation, and IMUs measure the angular velocity and acceleration of motion. The results of these measurements can significantly complement the errors and deficiencies of visual measurements.

1.2.2 Front-End: Visual Odometry (VO)

Visual Odometry (VO) focuses on the motion between adjacent images. As illustrated in Fig. 1.2, the primary challenge in VO is how to estimate camera motion based on two adjacent frames. However, images themselves are matrices composed of brightness and color. Direct motion estimation from the matrix level would be extremely challenging. Thus, current methods predominantly adopt the following approach: first, select representative points from the image (commonly referred to as visual feature points, with popular feature points including ORB [1] and SIFT [2], among others). These points remain consistent despite minor changes in the camera's viewpoint, allowing us to identify the same points in adjacent images. Then, based on these points, by obtaining the matching relationships between feature points in two adjacent frames, we further solve optimization problems to determine the motion relationship between frames. Finally, the absolute spatial positions of the spatial map points corresponding to the feature points are estimated

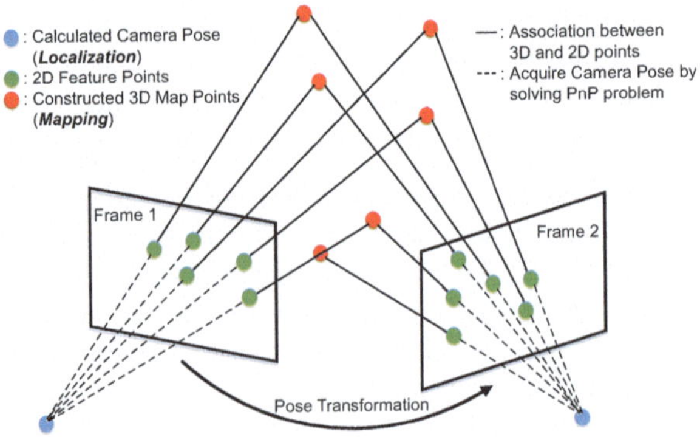

Fig. 1.2 Illustrations of visual odometry between two adjacent frames

through Triangulation [3], which is used to recover the three-dimensional structure of space.

After estimating the camera motion between two images, by stringing together the motion at adjacent moments, we construct the robot's motion trajectory, thereby addressing the localization problem. On the other hand, by calculating the positions of spatial points corresponding to each pixel based on the camera's position at each moment, we obtain a map. This raises the question: with VO, have we solved the SLAM problem? Certainly not. Estimating the trajectory solely through visual odometry will inevitably lead to accumulating drift. This is because VO only estimates the motion between two images, and each estimation carries a certain degree of error. More critically, due to the nature of odometry, errors from previous moments are transmitted to the next, causing the estimated trajectory to become inaccurate over time. To address the drift issue, two additional technologies are required: back-end optimization and loop closure detection.

1.2.3 Back-End: Non-linear Optimization

Back-end optimization primarily addresses the noise issues in the SLAM process. While we wish all data were accurate, in reality, even the most precise sensors carry some level of noise. Less expensive sensors have greater measurement errors, while more expensive ones have smaller errors, and some sensors are also affected by magnetic fields and temperatures. Therefore, besides solving the problem of *how to estimate camera motion from images*, we must also consider how much noise this estimation carries, how this noise is transmitted from one moment to the next, and how confident we are in our current estimation. The issues that back-end optimization needs to consider are how to estimate the state of the entire system from these noisy data, and how uncertain this state estimation is—known as Maximum-a-Posteriori (MAP) estimation [4]. The state to be estimated here includes both the robot's trajectory and the map.

Conversely, the visual odometry part is sometimes referred to as the *front-end*. In the SLAM framework, the front-end provides the back-end with the data to be optimized, along with their initial values. The back-end is responsible for the overall optimization process, often dealing only with data without having to concern itself with which sensor the data came from, as shown in Fig. 1.3. In visual SLAM, the front-end is more related to the field of computer vision research, such as feature extraction and matching of images, while the back-end mainly involves filtering and non-linear optimization algorithms. In a sense, the back-end optimization reflects the essence of the SLAM problem: the estimation of the uncertainty of both the moving entity itself and the spatial environment around it. To address the uncertainties in SLAM, we need state estimation theory to express the uncertainties in localization and mapping, and then use filters or non-linear optimization to estimate the state's mean and uncertainty (variance). Because the algorithms involved in back-end optimization are complexly expressed mathematically and

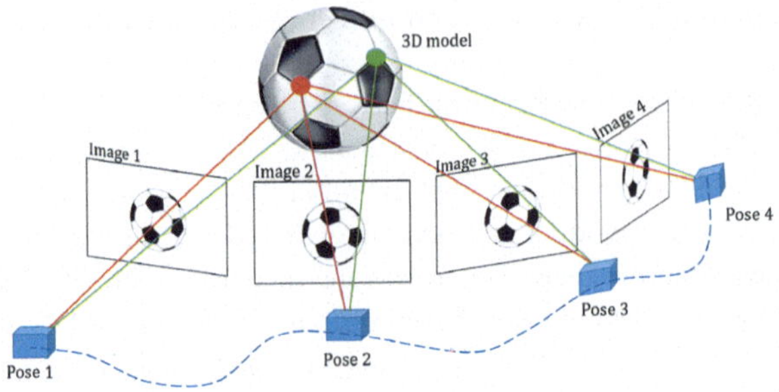

Fig. 1.3 Illustrations of back-end optimization

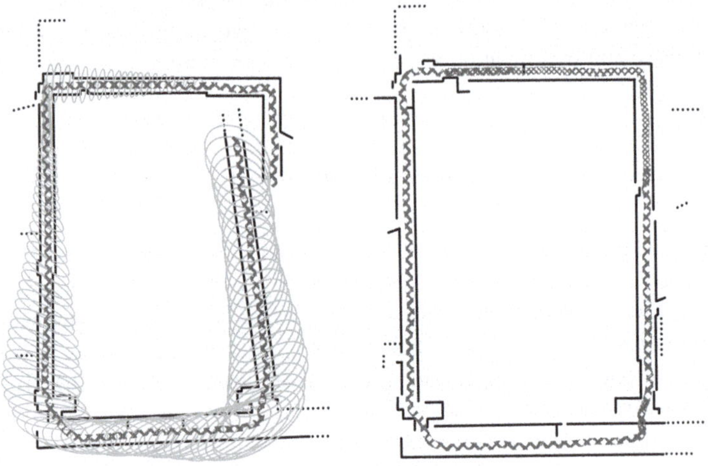

(a) Before loop closing (b) After loop closing

Fig. 1.4 Illustrations of the global map before and after loop closure detection.

there are numerous solution approaches, this book will introduce the corresponding optimization algorithms in later chapters in the context of specific applications.

1.2.4 Global: Loop Closure Detection

Loop Closure Detection, also known as Loop Closing, primarily addresses the issue of drift in position estimation over time. As illustrated in Fig. 1.4, assume that in reality, after a period of movement, a robot returns to its starting point. However, due to drift, its position estimation does not return to the starting point. What can be done? We think, if there were some way to let the robot know that it has returned to the starting point, or to identify the starting point, and then pull the position estimation back, we could eliminate the drift. This is the essence of loop closure detection.

Loop closure detection is closely related to both *localization* and *mapping*. In fact, we believe that the primary significance of a map is to let the robot know the places it has visited. To implement loop closure detection, we need the robot to have the ability to recognize scenes it has visited before. For example, we can judge the similarity between images to perform loop closure detection. This is similar to human perception. When we see two similar pictures, it is easy to recognize that they come from the same place. If loop closure detection is successful, it can significantly reduce cumulative errors. Thus, visual loop closure detection is essentially an algorithm for calculating the similarity of image data. The rich information in images significantly reduces the difficulty of correctly detecting loops. In current research, feature points can be directly utilized to match similarities. In recent years, researchers have also designed Bag of Words models [5] to statistically analyze the semantic information of pictures for robust loop closure detection.

After detecting a loop, we inform the back-end optimization algorithm that Point A and Point B are the same. Then, based on this new information, the back-end adjusts the trajectory and maps to match the results of loop closure detection. Thus, with sufficient and accurate loop closure detection, cumulative errors can be eliminated, resulting in a globally consistent trajectory and map.

1.2.5 Output: Mapping

Mapping refers to the process of constructing a map. A map is a description of the spatial environment, but this description is not fixed and depends on the application of visual SLAM. For a household robotic vacuum cleaner, which mainly moves in low planes, a two-dimensional map marking passable areas and obstacles is sufficient for its navigation within a certain range. However, for a camera that moves with six degrees of freedom, we need at least a three-dimensional map. Sometimes, we desire a beautifully reconstructed result, not just a set of spatial points, but also textured triangular patches. At other times, we do not care about the appearance of the map; we only need to know if "Point A to Point B is passable, while B to C is not." Sometimes, we do not need a map at all, or the map can be provided by others, such as vehicles that can often obtain pre-drawn local maps. When it comes to maps, we have too many ideas and needs. Compared to the previously mentioned visual odometry, loop closure detection, and back-end optimization, mapping does not have a fixed form or algorithm. A collection of 2D planar points can be called a map, a beautiful 3D model is also a map, and an image marked with cities, villages, railways, and rivers is also a map. The approximate appearance of the maps created by some SLAM systems can be seen in Fig. 1.5.

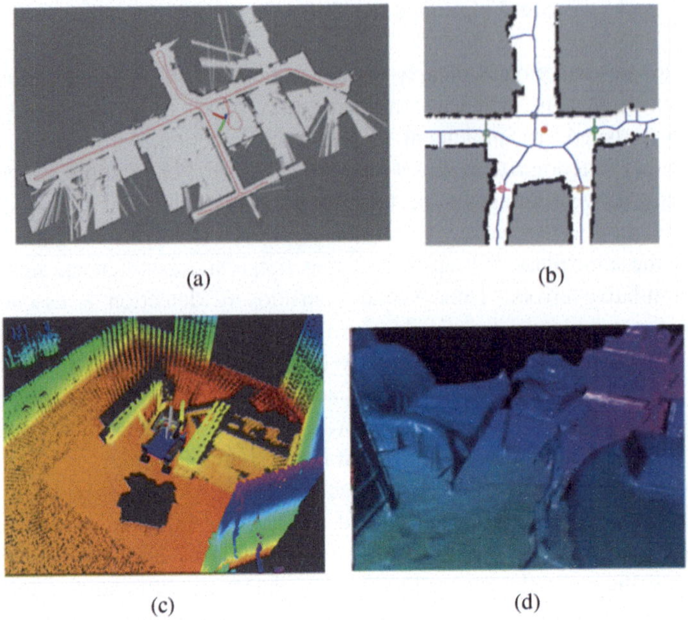

Fig. 1.5 Illustrations of different mapping outputs. (**a**) 2D grid map. (**b**) 2D topological map. (**c**) 3D point cloud map. (**d**) 3D mesh map

1.3 Challenges in Mobile Visual SLAM

Some classic visual SLAM systems, such as the ORB-SLAM series of works [6–8], LSD-SLAM [9], and VINS [10], have been deployed in the industry and have even become the core underlying algorithms for autonomous positioning and control systems in robots and drones [11]. However, running them on lightweight mobile devices still encounters a set of challenges that stem from both the operational and data management aspects of the technology.

Firstly, the computational overhead associated with the front-end VO and back-end optimization processes is quite high [12]. These essential components of the SLAM system involve complex calculations for feature extraction, matching, and optimizing the map and trajectory of the device based on incoming sensor data. Such intensive computations demand substantial processing power, which may exceed the capabilities of lightweight mobile devices, making it challenging to achieve real-time operation [13].

Additionally, the storage overhead for maps can be considerable. As the environment or the scene expands, the size of the map increases linearly, which can quickly consume the limited storage resources available on mobile devices. This issue is particularly pertinent for detailed 3D maps, which require storing extensive data on spatial points, textures, and potentially even semantic information to provide a comprehensive representation of the environment. The linear growth in map size with the expansion of the scene further compounds the problem, making it difficult to maintain and update large-scale maps on devices with constrained storage capacity [14].

In the subsequent chapters, we will demonstrate these challenges more intuitively through experiments and detailed analysis.

References

1. E. Rublee, V. Rabaud, K. Konolige, G. Bradski, ORB: An efficient alternative to sift or surf, in *Proceedings of the 2011 International Conference on Computer Vision (ICCV)* (IEEE, Piscataway, 2011), pp. 2564–2571
2. J. Wu, Z. Cui, V.S. Sheng, P. Zhao, D. Su, S. Gong, A comparative study of sift and its variants. Measur. Sci. Rev. **13**(3), 122–131 (2013)
3. R.I. Hartley, P. Sturm, Triangulation. Comput. Vision Image Understand. **68**(2), 146–157 (1997)
4. C. Cadena, L. Carlone, H. Carrillo et al., Past, present, and future of simultaneous localization and mapping: toward the robust-perception age. IEEE Trans. Rob. **32**, 1309–1332 (2016)
5. D. Gálvez-López, J.D. Tardos, Bags of binary words for fast place recognition in image sequences. IEEE Trans. Rob. **28**(5), 1188–1197 (2012)
6. R. Mur-Artal, J.M.M. Montiel, J.D. Tardos, ORB-SLAM: A versatile and accurate monocular SLAM system. IEEE Trans. Rob. **31**(5), 1147–1163 (2015)
7. R. Mur-Artal, J.D. Tardós, Orb-SLAM2: An open-source slam system for monocular, stereo, and RGB-D cameras. IEEE Trans. Rob. **33**(5), 1255–1262 (2017)
8. C. Campos, R. Elvira, J.J.G. Rodríguez, J.M. Montiel, J.D. Tardós, ORB-SLAM3: an accurate open-source library for visual, visual–inertial, and multimap slam. IEEE Trans. Rob. **37**(6), 1874–1890 (2021)
9. J. Engel, T. Schöps, D. Cremers, LSD-SLAM: Large-scale direct monocular SLAM, in *Proceedings of the European Conference on Computer Vision (ECCV)* (Springer, Berlin, 2014), pp. 834–849
10. T. Qin, P. Li, S. Shen, Vins-mono: A robust and versatile monocular visual-inertial state estimator. IEEE Trans. Rob. **34**(4), 1004–1020 (2018)
11. A.J. Ben Ali, Z.S. Hashemifar, K. Dantu, Edge-SLAM: Edge-assisted visual simultaneous localization and mapping, in *Proceedings of the ACM MobiSys* (2020)
12. E. Dong, J. Xu, C. Wu, Z. Yang, Y. Liu, Pair-navi: Peer to peer indoor navigation with mobile visual slam, in *Proceedings of the IEEE INFOCOM* (2019)
13. J. Xu, H. Cao, D. Li et al., Edge assisted mobile semantic visual SLAM, in *Proceedings of the IEEE INFOCOM* (2020)
14. J. Xu, H. Cao, Z. Yang et al., Swarmmap: Scaling up real-time collaborative visual slam at the edge, in *Proceedings of the USENIX NSDI* (2022), pp. 977–993

Chapter 2
Edge Computing in Mobile Visual Systems

Abstract To reduce the computational resource overhead of mobile devices when handling complex tasks, while simultaneously enhancing the system's real-time task inference, some studies have adopted **edge computing** solutions. This involves integrating relatively powerful edge servers, which are closer in network distance (typically within 3 hops), into the system design to provide computational support. In this chapter, we introduce the fundamental concepts and the overall technological framework of edge computing, providing readers with a better understanding of edge computing.

2.1 Understanding Edge Computing

As illustrated in Fig. 2.1, a typical edge-assisted system usually comprises three modules: the mobile end, the edge side, and the network, all of which collaborate through task scheduling to complete complex tasks that require significant computational power and real-time processing.

Specifically, the computationally limited mobile clients offload some of the relatively complex and resource-intensive sub-tasks (such as object recognition and environmental mapping in visual tasks) to edge servers for execution [1–3]. In contrast, they only need to focus on handling those latency-sensitive and application-critical lightweight sub-tasks. Meanwhile, after the edge side processes the complex sub-tasks, it transmits the intermediate results to the mobile end in a timely manner via network links, fulfilling the mobile end's dependence on these results for processing lightweight tasks.

Recent studies utilizing such edge-assisted schemes have achieved high-precision and real-time processing of complex tasks (such as object tracking and segmentation, autonomous localization, and environmental mapping) on lightweight mobile devices (such as commercial smartphones, small drones, autonomous robots, etc.) [4–6].

The interest in edge computing extends beyond academia, with the industry also actively promoting the implementation of edge computing in real-world production and daily life. For example, large industrial corporations like Siemens and Schneider

J. Xu et al., *Edge Assisted Mobile Visual SLAM*,
https://doi.org/10.1007/978-981-97-3573-0_2

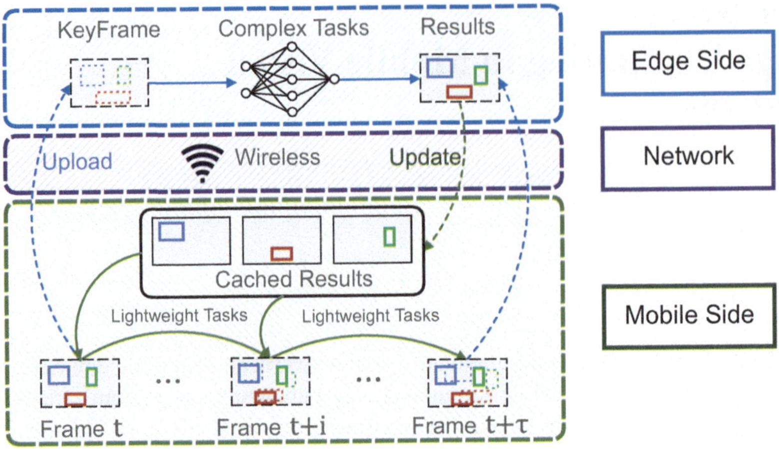

Fig. 2.1 Edge-assisted system architecture

Electric are using real-time digital modeling on edge devices to achieve *digital twins* of factories; hardware manufacturers like Intel and NVIDIA are also actively designing edge computing nodes tailored for specific tasks (such as the Nvidia Jetson TX series, AGX Xavier, etc.).

2.2 Enhancing Visual Tasks with Edge Computing

Although the introduction of a variety of sensors and advanced technologies has significantly enhanced the environmental perception and localization capabilities of mobile devices, the increasing complexity behind these advanced technologies (especially those related to computer vision, deep learning, or SLAM technologies) leads to a higher computational resource overhead. Consequently, mobile devices often struggle to perform tasks of self-motion estimation and environmental perception in real-time. However, for intelligent devices, localization and perception are merely foundational tasks; higher-level applications often rely on these results to perform more complex interactive tasks [6], highlighting the importance of completing localization and perception tasks in real-time with minimal computational resources.

In recent years, with the development of edge computing technology and the deployment of edge computing nodes, it has become possible for mobile devices to use edge servers to process some complex tasks (such as visual object detection and segmentation) in real-time.

Edge-assisted real-time task processing systems are primarily divided into three parts: the mobile end, the edge side, and the network. The overall technological framework has been introduced earlier. Some recent works have refined and upgraded this framework. For example, a series of works such as Glimpse [4], EAAR [7], Chameleon [8], and EdgeDuet [5] have proposed the *Track + Detect* framework, further improving the system's accuracy and real-time performance while saving the bandwidth required for transmission.

Taking EdgeDuet as an example, its *Track + Detect* system framework can be referred to in Fig. 2.1. On the mobile end, EdgeDuet uses lightweight target trackers, such as motion vectors [9] and Kernel Correction Filter (KCF), to "transfer" the cached detection results of previous keyframes to the current frame as the detection result for the current frame. This process is much faster than complex deep networks, and while sacrificing some accuracy, it can still meet the accuracy requirements as long as the content of the keyframe and the current frame is sufficiently similar. When the mobile end detects a significant difference between the current frame and the cached keyframe, it uploads this frame as a new keyframe to the edge server. The edge node then accurately locates the targets using relatively complex neural networks (such as YOLOv3 [10], ViT [11], etc.) and sends the results back to the mobile end to update the cache, providing tracking prior information for subsequent frames. Meanwhile, on the network, EdgeDuet does not simply upload the entire picture when transmitting keyframes but adopts a strategy of prioritized encoding for the Region of Interest (RoI) content, encoding important areas in high resolution and blurring non-critical areas to reduce the volume of data transmitted, thus enhancing the system's real-time performance.

Although edge-assisted real-time task processing systems have made significant progress in recent years, the framework and task scheduling strategies often need to be specially designed for specific applications, and related works mainly focus on applications in computer vision such as object detection and segmentation. Currently, there are no edge-assisted systems designed specifically for real-time SLAM systems on mobile devices.

References

1. Z. Chen, W. Hu, J. Wang et al., An empirical study of latency in an emerging class of edge computing applications for wearable cognitive assistance, in *Proceedings of the IEEE/ACM SEC* (2017)
2. L. Cheng, J. Wang, Vitrack: Efficient tracking on the edge for commodity video surveillance systems, in *Proceedings of the IEEE INFOCOM* (2018)
3. S. Jain, X. Zhang, Y. Zhou et al., Spatula: Efficient cross-camera video analytics on large camera networks, in *Proceedings of the IEEE/ACM SEC* (2020)
4. T.Y.-H. Chen, L. Ravindranath, S. Deng, P. Bahl, H. Balakrishnan, Glimpse: Continuous, real-time object recognition on mobile devices, in *Proceedings of the ACM Sensys* (2015)
5. X. Wang, Z. Yang, J. Wu, Y. Zhao, Z. Zhou, Edgeduet: Tiling small object detection for edge assisted autonomous mobile vision, in *Proceedings of the IEEE INFOCOM* (2021)
6. A.J. Ben Ali, Z.S. Hashemifar, K. Dantu, Edge-slam: Edge-assisted visual simultaneous localization and mapping, in *Proceedings of the ACM MobiSys* (2020)
7. L. Liu, H. Li, M. Gruteser, Edge assisted real-time object detection for mobile augmented reality, in *Proceedings of the ACM MobiCom* (2019)

8. J. Jiang, G. Ananthanarayanan, P. Bodik, S. Sen, I. Stoica, Chameleon: Scalable adaptation of video analytics, in *Proceedings of the ACM SIGCOMM* (2018)
9. J. Konrad, E. Dubois, Bayesian estimation of motion vector fields. IEEE Trans. Pattern Analy. Mach. Intell. **14**(09), 910–927 (1992)
10. J. Redmon, A. Farhadi, Yolov3: An incremental improvement (2018). arXiv preprint arXiv:1804.02767
11. Z. Liu, Y. Lin, Y. Cao et al., Swin transformer: Hierarchical vision transformer using shifted windows, in *Proceedings of the IEEE/CVF ICCV* (2021)

Part II
Edge-Assisted Visual SLAM: System Design Principle

Part II explores the system design and architecture that drive edge-assisted mobile Visual SLAM. It consists of two in-depth chapters.

Chapter 3 outlines the innovative architectural advancements, detailing the principles of system design, the integration of semantic SLAM with mobile devices, the architecture needed for real-time semantic analysis, and insights into adaptive system design for optimal performance.

Chapter 4 expands on scalability in multi-agent systems, presenting a new paradigm for scalable system design with SwarmMap and its innovative components. A case study on SwarmMap's application in oil-field exploration illustrates these concepts in action, offering a glimpse into practical deployment scenarios.

It is noteworthy that for ease of description in this and following parts, we use **EdgeSLAM** as an abbreviation for **Edge**-assisted mobile visual **SLAM**.

Chapter 3
EdgeSLAM 1.0: Architectural Innovations in Mobile Visual SLAM

Abstract In this chapter, we aim to exhibit the system architecture of edge-assisted mobile visual SLAM—that is, how visual SLAM can be accelerated with edge computing. Location services, critical for a myriad of applications, have garnered extensive attention from both the academic sphere and the industry sector. The rise of visual SLAM in domains such as robotics and autonomous vehicles marks a significant advancement. Nonetheless, its deployment on devices with limited resources is hampered by increasing computational demands. This chapter details the development, deployment, and assessment of edgeSLAM (a.k.a., EdgeSLAM 1.0), a cutting-edge service enabling semantic visual SLAM in real-time on mobile platforms. edgeSLAM capitalizes on advanced semantic segmentation techniques to refine both localization and mapping precision and mitigates the processing burden of demanding SLAM and segmentation tasks through strategic computation offloading. The pivotal breakthroughs of edgeSLAM encompass a streamlined offloading approach, a strategic data exchange protocol, and a dynamic task allocation framework. Comprehensive implementation on an edge framework alongside diverse mobile units (including two smartphone models and a developer board) supports edgeSLAM's efficacy. Rigorous testing across three datasets reveals that edgeSLAM sustains operation at a 35FPS rate while reaching up to 5 cm in localization accuracy, surpassing existing methodologies by over 15%. Furthermore, edgeSLAM's applicability is validated in pedestrian tracking and robotic navigation scenarios. To our knowledge, edgeSLAM stands as the pioneering endeavor to facilitate real-time semantic visual SLAM on mobile apparatus. We fully implement edgeSLAM on an edge server and different types of mobile devices (2 types of smartphones and a development board). Extensive experiments are conducted under 3 datasets, and the results show that edgeSLAM is able to run on mobile devices at 35FPS frame rate and achieves a 5 cm localization accuracy, outperforming existing solutions by more than 15%. We also demonstrate the usability of edgeSLAM through 2 case studies of pedestrian localization and robot navigation. To the best of our knowledge, edgeSLAM is the first real-time semantic visual SLAM for mobile devices.

J. Xu et al., *Edge Assisted Mobile Visual SLAM*,
https://doi.org/10.1007/978-981-97-3573-0_3

3.1 Introduction

Indoor positioning and navigation are crucial for numerous location-dependent services. The challenge in indoor settings is the poor performance of the Global Positioning System (GPS) due to significant signal loss and multipath effects [1–3], making it unsuitable for achieving the required precision. Despite rapid advancements in RF-based localization methods (e.g., Wi-Fi, RFID, Bluetooth) [4–6], their real-world application is hampered by either inadequate accuracy or prohibitive costs.

An emerging solution, vision-based positioning, specifically through visual SLAM, has gained traction [7–9]. It leverages camera-captured imagery alongside Inertial Measurement Unit (IMU) data to map surroundings and pinpoint the camera's locale with much higher precision compared to RF methods, achieving near 5 cm accuracy with minimal setup since cameras and IMUs are now standard in modern mobile devices [10].

However, even as mobile hardware and software advance (e.g., Samsung Galaxy S10's triple HD cameras enhance image and sensor accuracy), the computation capacity required for visual SLAM still outpaces what mobile devices can offer. Tests show that even streamlined algorithms like ORB-SLAM [7] struggle to operate in real-time (i.e., $\leq$15FPS) on contemporary smartphones (e.g., Galaxy S10, Google Pixel 2). Directly applying visual SLAM in such mobile contexts may not yield optimal results due to dynamic environmental variables (e.g., crowd movement in malls).

Yet, recent developments present new avenues for real-time mobile visual SLAM implementation:

1. The advent of edge computing alongside next-gen wireless technologies (5G [11], Wi-Fi 802.11ad [12]) offers promising prospects for managing compute-heavy tasks swiftly and locally, potentially accelerating visual SLAM processes through task offloading to edge servers.
2. Advances in computer vision (CV) methods (e.g., Mask-RCNN [13]) now permit high accuracy object recognition in images, opening the door for incorporating semantic insights from video captures to bolster SLAM efficiency.

However, realizing these intuitions is nontrivial and faces three significant challenges:

- **Task decomposition**. Simply transmitting all images back to an edge server is not feasible since it will introduces excessive bandwidth cost and transmission delays (details in Sect. 3.2.2). Partitioning both visual SLAM algorithm and semantic segmentation algorithm into unit tasks, however, is also nontrivial as both their functional units are tightly coupled. An improper partition may result in redundant data storage, exchange, and most importantly, system delay, which definitely in turn increases the algorithm latency.
- **Task cooperation**. The visual SLAM and semantic segmentation algorithms are generally regarded as two independent tasks, without the reuse of intermediate

results. However, to achieve both low-latency and high accuracy, it is beneficial to share the intermediate results between these two algorithms so that redundant computations can be eliminated or minimized.

- **Task scheduling**. The computation resource on mobile device and edge server is highly unbalanced. Meanwhile the wireless link between these two parts also varies from time to time. Task scheduling strategy should be adaptive to the dynamics in computation resource and wireless link quality.

In this chapter we present the design and implementation of edgeSLAM, a real-time **edge** assisted semantic visual **SLAM** service running on commercial mobile devices. edgeSLAM leverages the state-of-the-art semantic segmentation algorithm Mask-RCNN [13] to improve SLAM accuracy and speed up the SLAM and semantic segmentation algorithm by efficient computation offloading and data sharing, and adjust the offloading strategy automatically to adapt to the wireless link conditions.

To find out the optimal task decomposition strategy, we take the operation time, memory overhead and the transmission delay of each functional module into consideration and conduct extensive experiments to profile the performance of each module. We further analyze dependencies among these functional module and determine the "hourglass position" to decompose the Visual SLAM and object detection algorithm.

To minimize the latency introduced by redundant data transmission, we leverage the fact that the scene in most consecutive frames are similar. For example, the same billboard tends to appear in multiple consecutive frames. edgeSLAM avoids per-frame object segmentation operation and reuses the previous result from the last object segmentation until a significant frame change is detected on mobile devices.

To accommodate dynamic link conditions, we design a probing-optimizing strategy that first probe the network conditions and then leverages such information to optimize our task scheduling mechanism.

We fully prototype edgeSLAM's server side on an Ubuntu edge server and the client side on three different types of mobile devices including an Nvidia Jetson TX2 development board, a Samsung Galaxy S10 and an Apple iPhone X. The interface of edgeSLAM is shown in Fig. 3.1. Comprehensive experiments are carried out under different network conditions, such as Wi-Fi-5G, Wi-Fi-2.4G and Cellular-4G. We also examine edgeSLAM on two official datasets (TUM [14] and KITTI [15]) and a self-labeled dataset from three buildings. The results demonstrate that edgeSLAM could achieve average 35FPS frame rate with 5 cm localization accuracy and 2% relative mapping error in all scenarios, which outperforms existing systems by more than 10%. Our two case studies further demonstrate that edgeSLAM achieves outstanding performance in pedestrian localization task with average 9.6 cm accuracy and robot navigation task with 92.3% navigation success rate.

The key contributions are summarized as follows:

- We measured the operation time and memory overhead of each function module in visual SLAM and semantic segmentation and ascertained the optimal decoupling position and disassembly method.

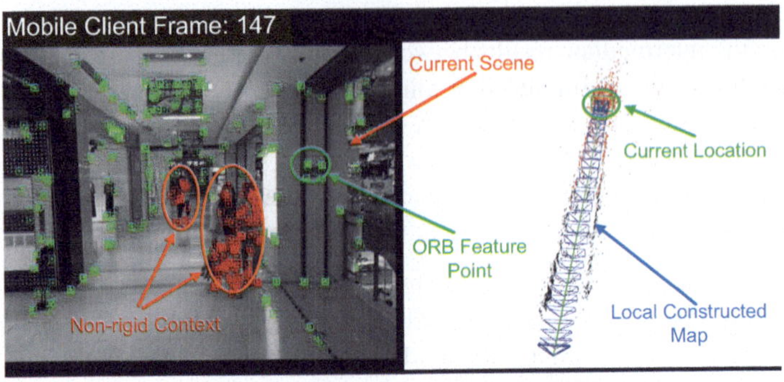

(a)

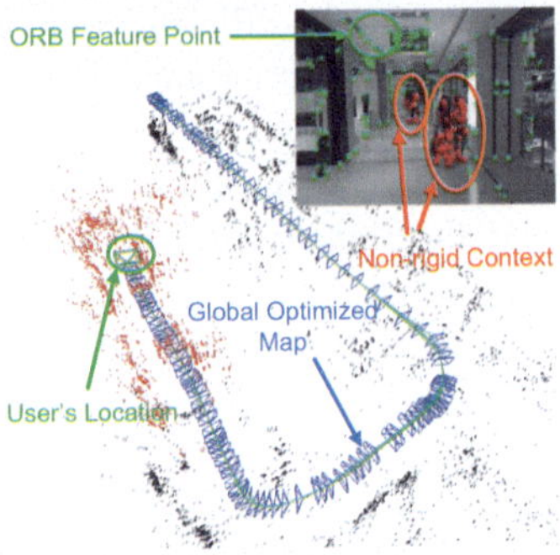

(b)

Fig. 3.1 Glimpse of edgeSLAM. (**a**) User's interface of edgeSLAM. Navigation instruction and local constructed map are displayed on mobile device. (**b**) Edge server's interface of edgeSLAM. Global map and user's location are optimized

- We designed the edge-assisted system architecture for visual SLAM and determined task assignment between mobile and edge, tackling the tightly coupled challenges behind visual SLAM's functional modules (Fig. 3.2). To the best of our knowledge, this is the first time that *mobile semantic visual SLAM* can work in real-time.
- We take network condition into consideration and adopt an adaptive method to dynamically adjust system parameters.
- We implement a complete edge-assisted real-time system on smartphones and extensively evaluate the performance. The results show that edgeSLAM achieves delightful results in all scenarios.

The rest of this chapter is organized as follows. We introduce the background and motivation of our work in Sect. 3.2. Followed by an overview of edgeSLAM in Sect. 3.3. Key strategies and techniques are presented in Sect. 3.4, followed by the dynamic design of self-adaptation strategy in Sect. 3.5. Related works are presented in Sect. 3.7. We finally conclude our work in Sect. 3.8.

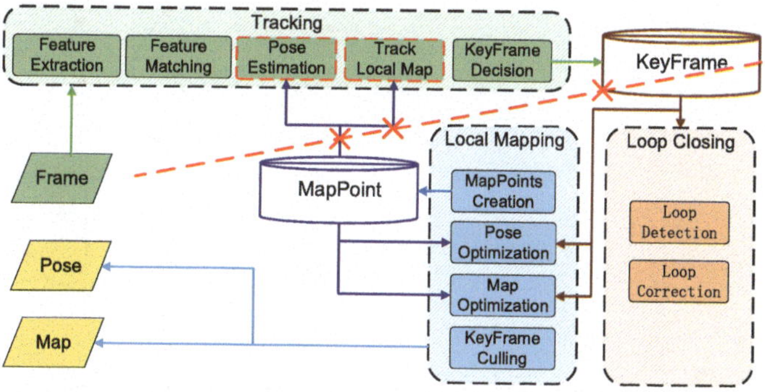

Fig. 3.2 Challenges of task assignment for semantic visual SLAM, where function units are tightly coupled

3.2 Understanding the Need for Edge Assistance

3.2.1 Semantic Visual SLAM

Semantic visual SLAM is structured around four principal modules: *tracking*, *local mapping*, *loop closing*, and *non-rigid context culling (NRCC)*, as depicted in Fig. 3.3. Here is a concise overview of each segment.

- The **Tracking** segment deduces the preliminary position of the recording camera from successive video frames. Upon receiving a new frame, it discerns 2D feature points, linking them to existing 3D map points. The matching process between 2D and 3D features, demonstrated in Fig. 3.4, yields an initial estimate of the camera's current stance.
- The **Local mapping** segment forms fresh 3D map points by triangulating between adjacent frames [16]. By addressing a Bundle Adjustment issue, it refines the camera's pose. As more images are captured, this segment compiles a trajectory of camera positions, a catalog of 3D landmarks, and relevant keyframes.

Fig. 3.3 System architecture of semantic visual SLAM

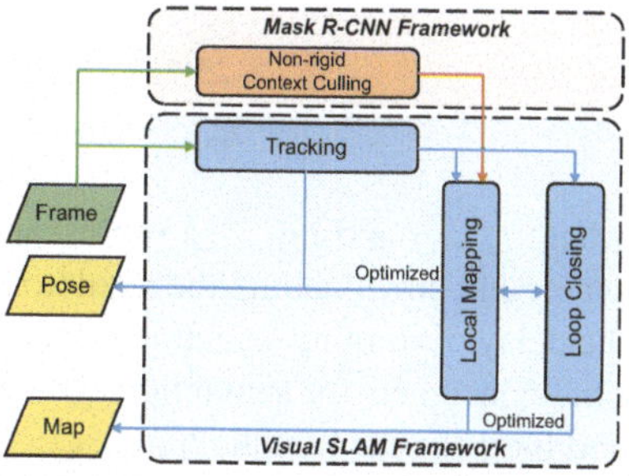

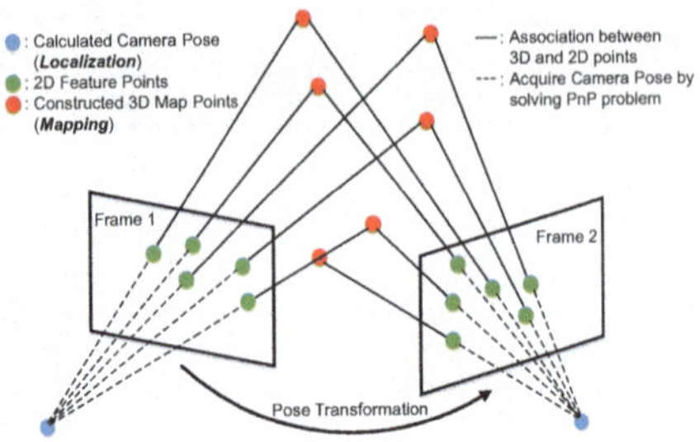

Fig. 3.4 Illustration of visual SLAM

- The **Loop closing** segment evaluates the frame's features against keyframes to ascertain resemblances. Recognizing a keyframe akin to the current frame activates adjustments to the camera pose and map optimization based on the identified match.
- The **NRCC** segment identifies temporary objects (like pedestrians) in video frames to mitigate their impact on both localization and mapping.

3.2.2 Latency and Accuracy Analysis

Running semantic visual SLAM on mobile devices is challenging due to its extensive computation overhead. For example, the classical visual SLAM algorithm ORB-SLAM [7] along works in a rate of only 10FPSwhen running on a Samsung Galaxy S10 smartphone, far away from the real-time requirement (≥30 FPS [17, 18]). This frame rate will drop further when the semantic segmentation algorithm (e.g., Mask-RCNN) runs in parallel to SLAM on mobile devices. Offloading the entire computation to the powerful edge server, on the other hand, may cause considerable latency. To better understand this transmission latency and its impact on the localization and object segmentation performance, we conduct experiments detailed below.

Latency Analysis We model the end-to-end latency (from capturing a video frame until we obtain a camera pose) of a semantic visual SLAM solution as follows:

$$
\begin{aligned}
t_{\text{e2e}} &= t_{\text{upload}} + t_{\text{infer}} + t_{\text{download}} \\
t_{\text{infer}} &= \max(t_{\text{NRCC}}, t_{\text{T}}) + t_{\text{LM}} + t_{\text{LC}},
\end{aligned}
\tag{3.1}
$$

where t_{upload} and t_{download} represent the delay for uploading an image to an edge server and downloading the result to the mobile device, respectively. t_{infer} represents the delay of running semantic visual SLAM on an edge server, which consists of the time consumed for semantic segmentation t_{NRCC}, tracking t_{T}, local mapping t_{LM}, and loop closing module t_{LC}, respectively.

We measure the latency of each functional module in various wireless link connections, and show the result in Fig. 3.5a. From the result we can see that

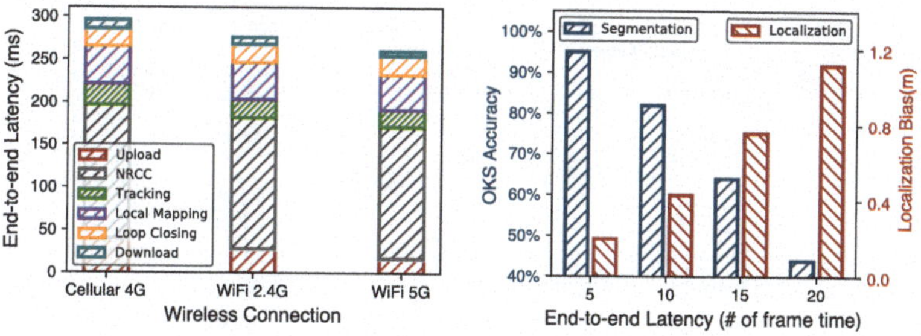

(a) (b)

Fig. 3.5 Accuracy and latency analysis. (**a**) Localization and segmentation accuracy with respect to different end-to-end latency. (**b**) End-to-end latency with different wireless connection link

offloading the entire computation to an edge server introduces significant latency (280 ms on Cellular 4G link, 270 ms on Wi-Fi 2.4G link, and 260 ms on Wi-Fi 5G link, respectively). In other words, we deserve only 4FPS over all three wireless links. While SLAM and semantic segmentation can be further accelerated by leveraging more advanced edge server, the uploading and downloading latency, on the other hand, still takes 48 ms, which set an upper bound of the achievable frame rate (20FPS given the ignorable inference latency).

Accuracy Analysis We further conduct experiment to understand the impact of different delays on the localization accuracy and semantic segmentation accuracy. In this trail of experiments, we record the 3-D position of the camera reported by the visual SLAM algorithm and calculate the distance between the inference result and the ground truth (measured by a laser ranger) using bias Euler-distance metric. We also use the Object Keypoint Similarity (OKS) [19] metric to measure the accuracy of keypoints in each group in the object keypoint segmentation task.

Figure 3.5b shows the localization and semantic segmentation accuracy as a function of end-to-end delay. We observe that the localization error maintains in a very low level when the end-to-end latency is low (i.e., five frames delay). The localization error then jumps to over 1 m as we increase the end-to-end delay to 20 frames (about 666 ms). The segmentation accuracy shows the similar trend: it decreases from over 90% accuracy to less than 15% accuracy as we increase the end-to-end delay from 5 frames to 20 frames. This result clearly demonstrates that the end-to-end latency has a significant impact on both two algorithms' performance.

3.3 System Architecture for Edge-Assisted SLAM

Addressing the identified challenges, we introduce edgeSLAM, a cutting-edge semantic visual SLAM framework designed for mobile applications that concurrently delivers precision and real-time performance through edge computing

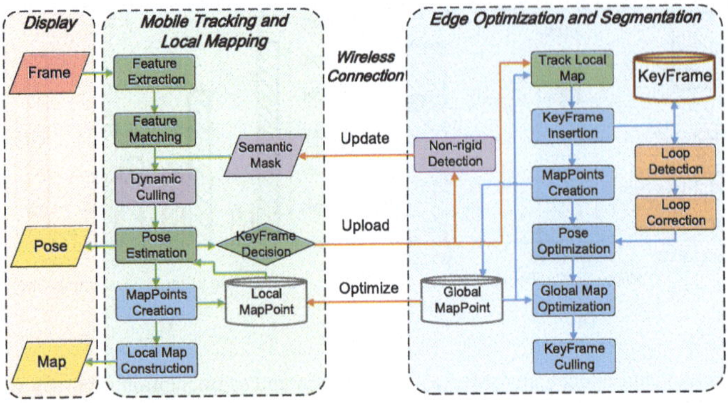

Fig. 3.6 edgeSLAM architecture

resources. edgeSLAM strategically segments the integrated visual SLAM workflow into dual pathways to mask offloading latency for the semantic SLAM tasks. Illustrated in Fig. 3.6, these bifurcated processes are interlinked via a wireless connection, encapsulating the *Mobile Tracking and Local Mapping* on mobile units (such as smartphones for individuals or development boards for robotic applications) and the *Edge Optimization and Segmentation* at the edge infrastructure. The initial pathway undertakes the camera's pose tracking and local map generation, whereas the latter concentrates on pixel-level frame segmentation, refining initial pose estimations, and global map augmentation. Interactions between the two realms are initiated upon the selection of keyframes by the mobile pathway, which are then transmitted to the edge. Subsequently, the edge server dispatches the enhanced pose and map data back to the mobile domain alongside the keyframe's segmentation details, facilitating semantic mask transfer across video sequences. Upon receipt of optimization data, the mobile client adjusts and integrates the current camera pose and local map into the live scene, executing this process in tandem with the mobile pipeline to maintain continuous, real-time tracking and mapping display.

Summarily, edgeSLAM stands out as a groundbreaking real-time, mobile semantic visual SLAM solution, distinguished by its:

- Modular decomposition of visual SLAM into granular components, redistributed between mobile and edge to facilitate real-time operation on the mobile platform.
- Innovative *Parallel Local Tracking and Global Optimization* methodology that conceals the intensive optimization tasks behind ongoing local tracking and mapping activities, thereby preserving accuracy.
- Implementation of a *Semantic Mask Transfer Strategy*, enabling the application of pixel-level semantic data from keyframes to adjacent frames, significantly diminishing inference delays and enabling semantic segmentation on mobile platforms.

The subsequent sections delve into the intricacies of these innovative approaches.

3.4 System Design for Edge-Assisted SLAM

3.4.1 *Decoupling and Task Re-assignment of Visual SLAM*

In the ORB-SLAM framework, significant computational delays are primarily due to the operations within the *Tracking* thread and optimization tasks in the *Local Mapping* and *Loop Closing* threads. For mobile devices, the duration needed for both pose estimation and map-point generation exceeds 33.3 ms, rendering the system incapable of achieving a frame rate above 30FPS. These processes are dependent on the Map-Point and KeyFrame databases for comprehensive optimization, as illustrated in Fig. 3.7 through a breakdown of ORB-SLAM's execution time.

To mitigate these issues, we propose a strategy based on task separation and redistribution that adheres to three core guidelines: Firstly, the mobile device should handle its designated tasks within real-time constraints, specifically in under 33.3 ms. Secondly, modules requiring interactions with the KeyFrame or Map-Point databases for advanced optimization are better suited for execution on the edge server to avoid the latency spikes caused by data dependencies. Lastly, to align with traditional visual SLAM systems, both the tasks of localization (monitoring the camera's position) and map creation (generating map points) must occur on the mobile device.

edgeSLAM's task distribution strategy meets these criteria effectively. As depicted in Fig. 3.6, the mobile unit preliminarily estimates the camera's approximate pose by extracting and matching feature points, also keeping a smaller, local map-point database for immediate map construction. This local database typically comprises about 10% of the global database's size. Concurrently, the edge server handles the demanding and computation-intensive optimization tasks, which include refining poses and maps and managing the extensive global databases for map points and keyframes. Upon receiving a keyframe, the edge will dispatch the

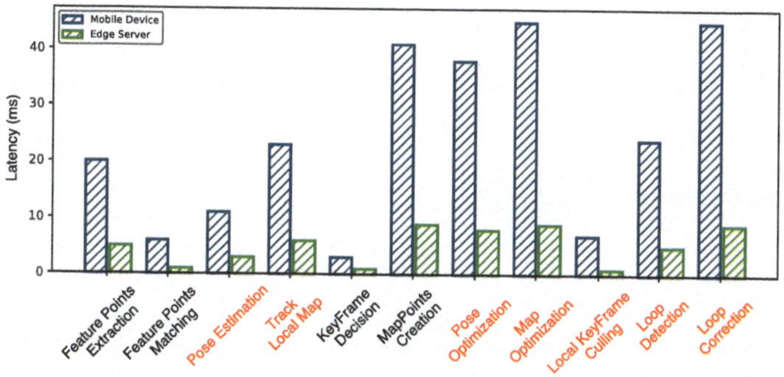

Fig. 3.7 Operation latency of each function unit in visual SLAM

refined pose and the refreshed map-point data back to the mobile device, which then adjusts for any cumulative errors in tracking and mapping.

3.4.2 Parallel Local Tracking and Global Optimization

The server performs complete optimization-related modules in parallel with the *Mobile tracking* process on the mobile client. We further design a method to update local constructed map on client with server optimized map, as illustrated in Fig. 3.8. Transferring the enormous volume of the entire optimized global map will increase the transmission latency as well as data serialization time dramatically. Thus, we monitor the modification of the Map-Point database updated by the keyframe and generate a map slice, which records the creation, deletion, and calibration of each map-point by their unique ID as a primary key. Then, we send the map slice rather than entire global optimized map to client for updating. The updating procedure is also executed synchronously with mobile tracking process. By doing so, we can achieve incremental client map optimization, meanwhile decrease the transmission latency and serialization/deserialization time.

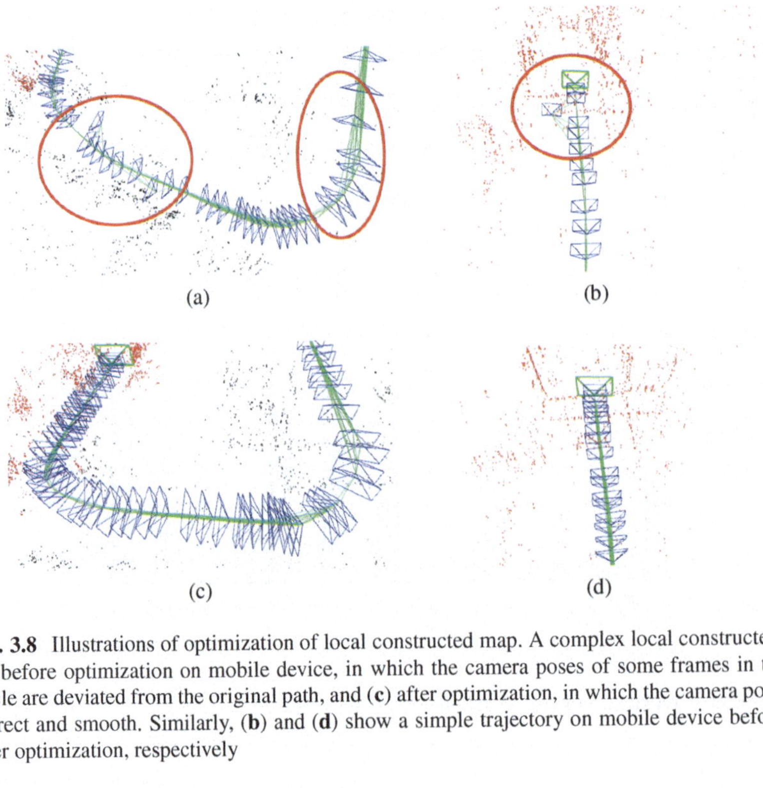

Fig. 3.8 Illustrations of optimization of local constructed map. A complex local constructed map (**a**) before optimization on mobile device, in which the camera poses of some frames in the red circle are deviated from the original path, and (**c**) after optimization, in which the camera poses are correct and smooth. Similarly, (**b**) and (**d**) show a simple trajectory on mobile device before and after optimization, respectively

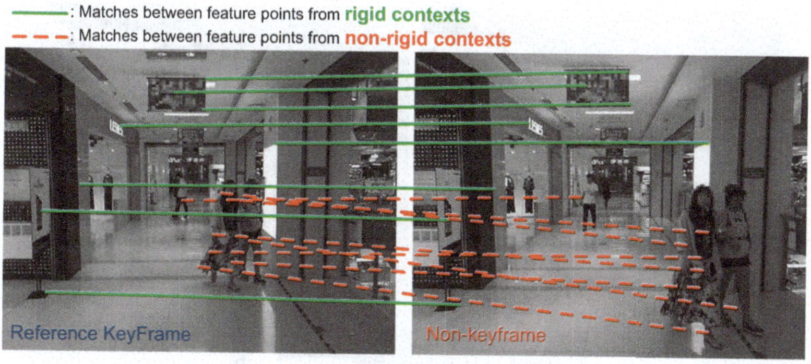

Fig. 3.9 Illustration of semantic mask transfer strategy

3.4.3 Semantic Mask Transfer Strategy

Same as many recent works, we adapt Mask R-CNN for NRCC. Mask R-CNN is a famous framework for instance segmentation. It aims to separate different instances in an image via a segmentation mask for each instance. However, despite high accuracy, it is time-consuming and resource-awareness. As shown in Fig. 3.5a, it results in more than 150 ms latency to infer a video frame, which is the bottle-neck of semantic visual SLAM.

In edgeSLAM, we design a semantic mask transfer strategy, which can obtain the segmentation information on mobile in real-time. More specifically, as illustrated in Fig. 3.9, only keyframe selected by mobile pipeline will be uploaded to the edge cloud and calculated semantic mask. As for non-keyframe, the pixel-level semantic information will be transferred from the latest keyframe according to feature points matching. The rationale behind the transfer strategy is: First, feature points matched between two consecutive frames enjoy larger probability of belonging to the same object categories according to the matching algorithm (DBoW2) in ORB-SLAM. Second, a user (or robot) will move less than 4 cm during the 33.3 ms (30FPS) time-window, assuming that a typical walking speed is 1 m/s. Therefore the potential scene changes in two consecutive frames will not be exceedingly significant. Only inferring keyframes whose scene have changed a lot compared with the previous keyframe, is a more effective way. Semantic information will be updated once the mobile client receives a fresh semantic mask.

3.5 Self-Adaptation Strategy Design

For better adaptation to diverse environments, we considered the mobile pipeline of edgeSLAM and found that there are several parameters available for adjustment in the keyframe selection function. By adjusting these parameters dynamically,

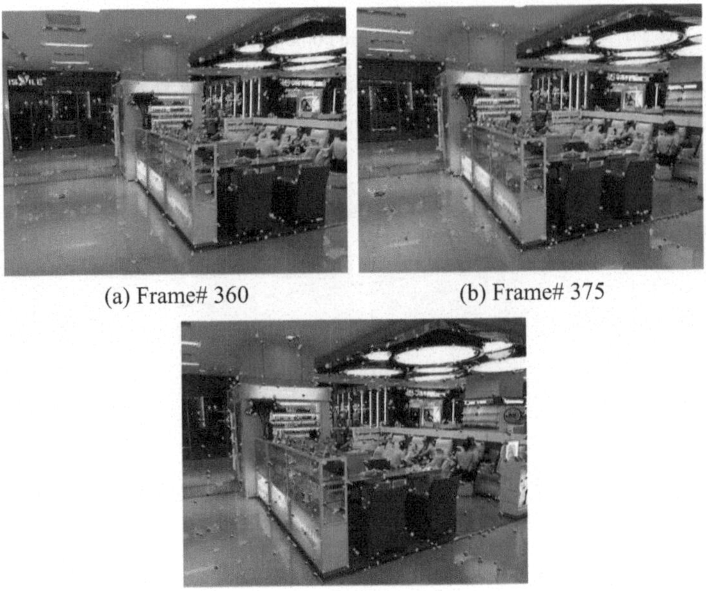

(a) Frame# 360 (b) Frame# 375

(c) Frame# 390

Fig. 3.10 Illustration of keyframe selection strategy. (**a**) A reference Keyframe. (**b**) A non-keyframe, more than 90% of the extracted feature points are matched with (**a**) previous Keyframe. (**c**) A new keyframe, whose associating rate with (**a**) is less than 80%

edgeSLAM will achieve relatively optimal results under various computing power restrictions and network conditions.

The basis of keyframe selection is the extraction of feature points, which leverage Gaussian Pyramid to ensure the scale invariance. Different Gaussian Pyramid layers (n_l, nlevels) will lead to diverse ranges of corner point search. With more layers, the computing power consumption will also increase. Thus, the corresponding adjustments is required when the end device has limited computing power.

After extraction of feature points, we design following rules in edgeSLAM as principles for keyframe selection:

1. More than $f_{\min}$ (mMinFrames, 20 by default) frames have passed from the last keyframe selection on mobile device, or the number of keyframes in the map is less than $f_{\min}$ frames.
2. Less than $f_{\max}$ (mMaxFrames, 60 by default) frames have passed from the last optimization for local constructed map.
3. The ratio of extracted feature points in current frame compared with previous keyframe should be at least α (0.4 by default).
4. The ratio of matched feature points in current frame compared with previous keyframe cannot exceed β (0.9 by default).

Table 3.1 Different configuration tuples in self-adaptation strategy

Configuration#	f_{min}	f_{max}	α	β	n_l	n_m	N (ms)
1	10	40	0.3	0.9	8	0.8	(0, 10]
2	20	50	0.3	0.9	6	0.8	(10, 15]
3	20	60	0.4	0.9	6	0.8	(15, 35]
4	30	70	0.5	0.8	4	0.7	(35, 50]
5	40	80	0.5	0.7	4	0.6	(50, 300)

Condition 1 ensures a good optimization and condition 3 a good tracking. They enhance the quality of selected keyframes. Condition 2 and 4 reduce the redundancy of keyframes and ensure representativeness of them. An example of the keyframe selection strategy is shown in Fig. 3.10.

Considering the principles above, we can appropriately increase f_{max} and f_{min} or reduce α and β under poor network conditions.

Apart from variables we mentioned above, non maximum suppression parameter n_m in Mask R-CNN framework is also an essential parameter, which makes trade-off between segmentation accuracy and latency. We will also adjust it to adapt different requirements in diverse environments.

Similar to DeepDecision [20], our self-adaptation strategy takes environment measurements (network bandwidth B and network latency L) as inputs and defines the network condition N as:

$$N = L + \frac{M}{B}, \tag{3.2}$$

where M is the data size for each frame (1280 $\times$ 720 pixels in edgeSLAM). Furthermore, edgeSLAM outputs the optimization strategy tuple $(f_{min}, f_{max}, \alpha, \beta, n_l, n_m)$. Specifically, we set five configuration tuples to make edgeSLAM adapt to various network conditions, which can be seen in Table 3.1.

3.6 System Implementation and Evaluation

3.6.1 Implementation

Client The implementation edgeSLAM follows the system workflow in Fig. 3.6. We implement the client part of edgeSLAM on mobile devices (Nvidia Jetson TX2, Apple iPad mini 5, iPhone X, and Galaxy S10). To begin with, we continuously capture video frame by camera on devices using OpenCV [21] API and JetPack Camera API and feed it to *mobile tracking and local mapping* thread. `ORBextractor()` function will extract ORB feature points. Meanwhile, CUDA (only on the development board) will accelerate the process [22]. This process is followed by `featurePointsMatching()` function to calculate associations between extracted feature points. Furthermore, `featureCulling()` takes over the procedure, which leverages semantic mask to ensure the rigidity of environment. Keyframe decision function `isKeyFrame()` will be finally performed according to principles described in Sect. 3.5. It determines whether or not the new created frame is a keyframe. Keyframe will be uploaded to edge server. Meanwhile, `poseEstimation()` and `localMapGeneration()` will estimate the coarse-grained camera pose and construct a map. Once client receives a map slice sent from edge server, `Update()` function will optimize pose and map.

Server The server's SLAM program is developed on *ORB-SLAM*[23], and we develop the system visualization upon the visualization of *ORB-SLAM*, with some modifications via OpenCV. The server is equipped with Intel(R) Xeon(R) CPU E5-2620v4 of 2.10 GHz main frequency and 256G RAM, running the Ubuntu 16.0.4 operating system. For Mask R-CNN, the GPU we use is TITAN V with cuda version 9.1.85 and cudnn-7.05. We apply our Mask R-CNN models with the ResNet-FPN-50 backbone and the network parameters are pretrained on COCO image Dataset [24]. The Mask R-CNN code is implemented in python-3.6.5 with pytorch-0.4.0 [25].

Remote Data Interaction The Remote Procedure Call (RPC) refers to the form of inter-process communication (IPC), however different processes have distinct address space. The RPC model helps procedures executed in local invoke remote messages and functions. It is perfectly suitable for Keyframe uploading and optimized map slice updating in edgeSLAM. By applying a widely-used RPC framework gRPC [26] developed by Google, we achieve efficient data transmission and remote function invocation. We also leverage the boost serialization library [27], which can be used to reconstitute an equivalent structure in another program context. Moreover, the strategy can also help to encode the structure to binary stream and embed it in Protobuf [28] package as the RPC arguments, thus, saving network bandwidth and accelerating the data transmission rate.

3.6.2 Experiment Setup

Our system underwent rigorous testing on two renowned SLAM benchmarks: TUM [14] and KITTI [15], to ascertain its efficacy. The TUM dataset, known for its sequences complemented by precise motion capture system-derived ground truth, serves as an ideal platform for assessing camera localization precision. Meanwhile, the KITTI odometry benchmark, with its 11 sequences captured in residential settings and validated against GPS and Velodyne laser scans, presents a formidable challenge for monocular vision systems due to rapid rotational movements and densely foliated regions complicating data matching.

Our evaluation focused on the localization precision of edgeSLAM across 16 indoor sequences from the TUM RGB-Monocular benchmark and the camera pose tracking and map construction efficiency across five KITTI sequences. An Nvidia Jetson TX2 development board and a conventional desktop computer were employed as the mobile and edge server units, respectively, for these tests. The setups used are as detailed previously. In these trials, images from the datasets were processed at a consistent 30 FPS by the mobile device, with edgeSLAM operating seamlessly at the acquisition rate.

Additionally, to thoroughly gauge edgeSLAM's edge-assisted framework's impact, we drew comparisons with **ORB-SLAM** and **Mask-SLAM**, the latter being employed in recent studies [18, 29] for semantic information integration and NRCC application in video analysis. It is noteworthy that the baseline **ORB-SLAM**, our edgeSLAM's precursor devoid of edge support and NRCC, alongside **Mask-SLAM**, fails to deliver real-time performance on mobile units. Consequently, in these instances, computational duties are entirely transitioned to the edge server upon image capture by the mobile device.

3.6.3 Performance Comparison

3.6.3.1 Localization Accuracy in TUM Dataset

We first examine the localization accuracy of edgeSLAM. Figure 3.11a depicts the performance of the proposed edgeSLAM as well as two other comparative systems in indoor localization scenarios. As shown, the average localization accuracy of edgeSLAM, ORB-SLAM, and Mask-SLAM is 0.83 cm, 1.24 cm, and 1.97 cm, respectively. edgeSLAM outperforms the other two approaches by more than 30%. Moreover, our edgeSLAM can work on the mobile device in real-time, which is the unique advantage among these approaches. The delightful result comes from the low end-to-end latency, which is ensured in edgeSLAM by edge-assisted method and task assignment strategy.

3.6.3.2 Pose Tracking Accuracy in KITTI Dataset

We further examine the camera pose tracking accuracy of edgeSLAM in KITTI camera rotation dataset. We calculate the Relative Rotation Error (RRE, an essential evaluation indicator in KITTI dataset) of the tracking result compared with the ground truth. The performance of edgeSLAM as well as two comparative methods are depicted in Fig. 3.11b. The tracking accumulative bias of edgeSLAM is within 0.22° for one meter length traces, which outperforms ORB-SLAM and Mask-SLAM by 5.1 and 17.6%.

3.6.3.3 Mapping Precision in KITTI Dataset

Finally, we evaluate the mapping precision of edgeSLAM in KITTI Dataset. We calculate the Relative Translation Error (RTE, another fundamental evaluation

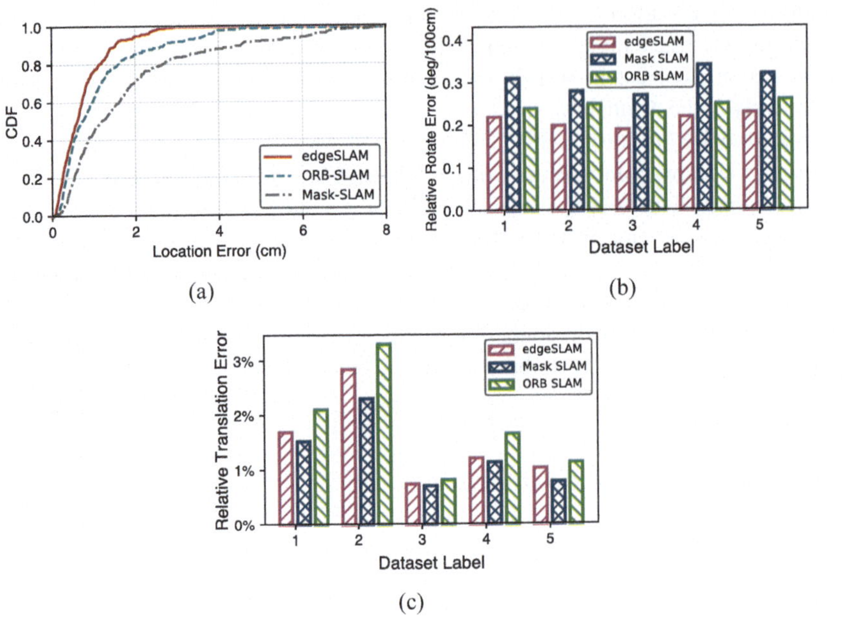

Fig. 3.11 System performance comparison. (**a**) Localization accuracy on TUM dataset. (**b**) Tracking accuracy on KITTI dataset. (**c**) Mapping accuracy on KITTI dataset

indicator in KITTI dataset) of the constructed map compared with ground truth. In this experiment, we compared the global optimized map stored in edge server in edgeSLAM with maps constructed by other two systems. As shown in Fig. 3.11c, in all five image sequences, edgeSLAM outperforms ORB-SLAM by more than 12% and the performance gap between edgeSLAM and Mask-SLAM is within 10%, which demonstrate that edgeSLAM achieves competitive performance.

In summary, real-time edgeSLAM achieves enhanced localization accuracy and competitive mapping performance with state-of-the-art offline frameworks in all evaluations. The grateful performance comes from not only the efficient design of edge-assisted method but also the adaption of semantic mask transfer strategy.

3.6.4 End-to-End Latency

Our system is able to achieve an end-to-end latency within 33.3 ms inter-frame time at 30FPS to ensure a smooth localization and mapping experience. To validate this, we run edgeSLAM under different network connections and calculate the average end-to-end latency for each frame in Fig. 3.12. The red dashed line in the figure is the 33.3 ms deadline for 30FPS SLAM devices. We can find that our system is able to finish the entire task within 33.3 ms under the majority wireless connection

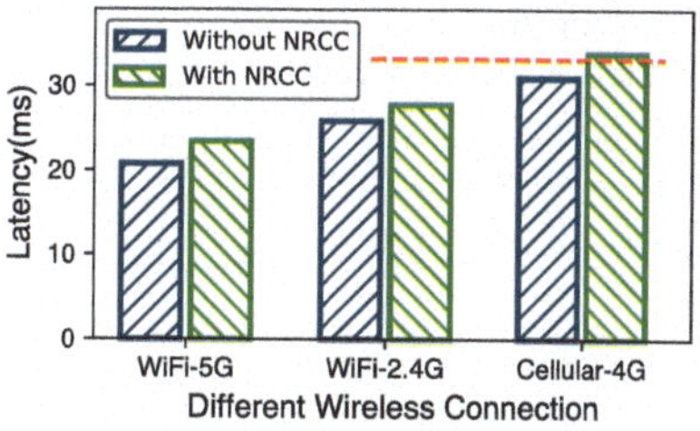

Fig. 3.12 End-to-end latency comparison

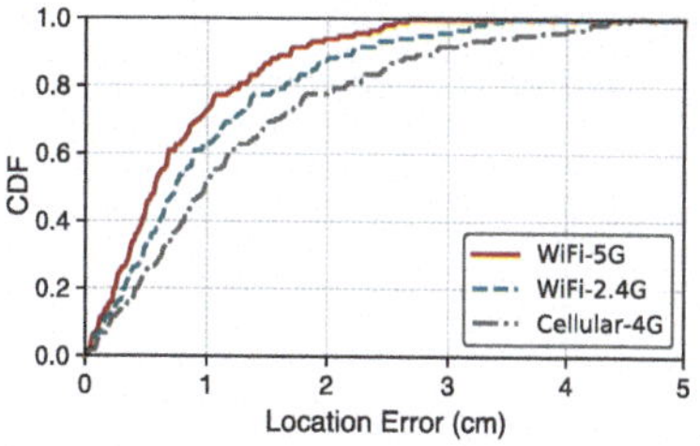

Fig. 3.13 Localization accuracy comparison

conditions (34.1 ms under 4G wireless connection with NRCC). Moreover, the average frame latency increases merely <4 ms if NRCC is involved in edgeSLAM, which is the bottle-neck in most systems as aforementioned.

Furthermore, we evaluate the robustness of edgeSLAM under different network conditions. As shown in Fig. 3.13, edgeSLAM can reach an average accuracy of 0.7 cm, 0.9 cm, and 1.3 cm under different wireless connections, respectively. The drift of accuracy influenced by network condition is within 0.5 cm.

3.6.5 Impact of Self-Adaptation Strategy

To demonstrate the effectiveness of designed self-adaptation strategy. We evaluate the average frame latency and system localization accuracy with and without leveraging the strategy. edgeSLAM will upload each frame once mobile client receives the optimization result of previous uploaded frame if without self-adaptation strategy. As shown in Fig. 3.14, after bringing in the adaptation strategy, the average frame latency decreases more than 5 ms, especially >10 ms under Cellular-4G wireless connection. The result reflects that the designed self-adaptation strategy plays an important role in ensuring edgeSLAM runs in real-time, especially under negative network condition. Figure 3.15 shows the performance of edgeSLAM with and without adaptation strategy. As seen, the average localization accuracy are 1.03 cm and 0.93 m, respectively. The precision difference is within 10%. In a nutshell, above results show that the leverage of self-adaptation strategy effectively decreases the end-to-end latency while yields similar performance.

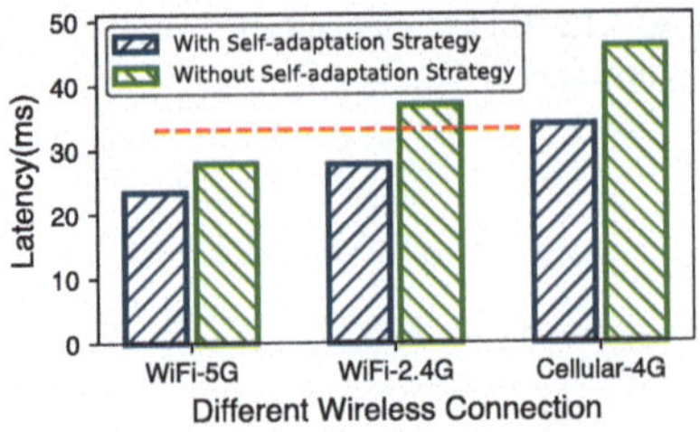

Fig. 3.14 Impact of self-adaptation on latency

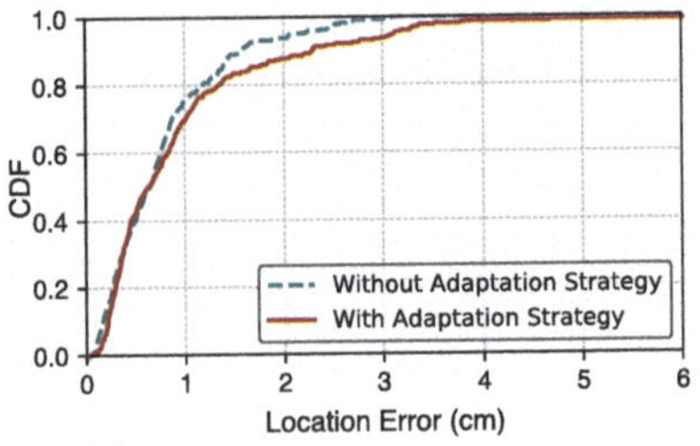

Fig. 3.15 Impact of self-adaptation on accuracy

3.6.6 Case Study

3.6.6.1 Pedestrian Localization and Tracking

We conducted experiments in a laboratory and first floor of a shopping mall. These two areas have different floor layouts, diverse wireless environments, and distinct user behavior patterns. In particular, the crowded shopping mall is the most dynamic. While there are a reasonable number of users in the laboratory most of the time.

Setup In this case, client side of edgeSLAM is implemented on an iOS platform (Apple iPhone X). The server we use is a Lenovo IdeaPad-Y700 with i7-6700HQ CPU of 2.6 GHz main frequency and 16G RAM, running the Ubuntu 16.0.4 operating system. For Mask R-CNN, the GPU we used is TITAN V with Cuda version 9.1.85 and cudnn-7.05.

Ground Truth Acquisition In order to obtain the ground truth, which is the accurate location of pedestrian, we recruited a volunteer to watch surveillance videos, artificially differentiate and track pedestrian and manually mark their locations on the 2D indoor map.

Localization Result Analysis Figure 3.16 shows the performance of edgeSLAM for pedestrian localization in different areas. As seen, edgeSLAM yields an average accuracy of 7.6 cm in the laboratory, and 9.9 cm in the shopping mall. The corresponding 95th percentile location errors in these two buildings are 9.9 cm, 11.4 cm, respectively. The result shows that edgeSLAM can locate a pedestrian at fine-grained in real-time (40FPS in this case study), outperforming state-of-the-art RF-based and vision-based localization systems. Moreover, edgeSLAM yields

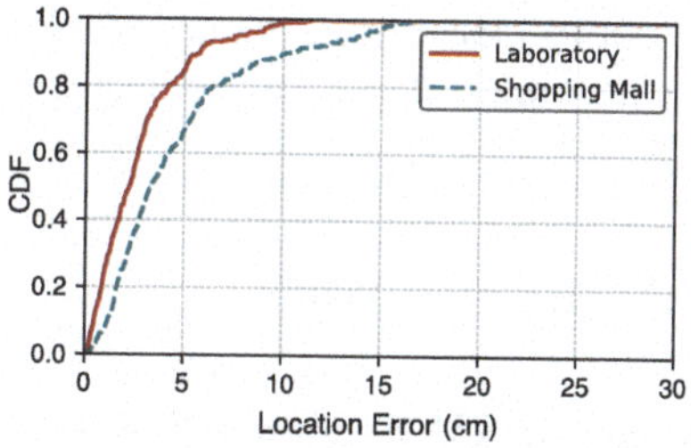

Fig. 3.16 Pedestrian tracking accuracy

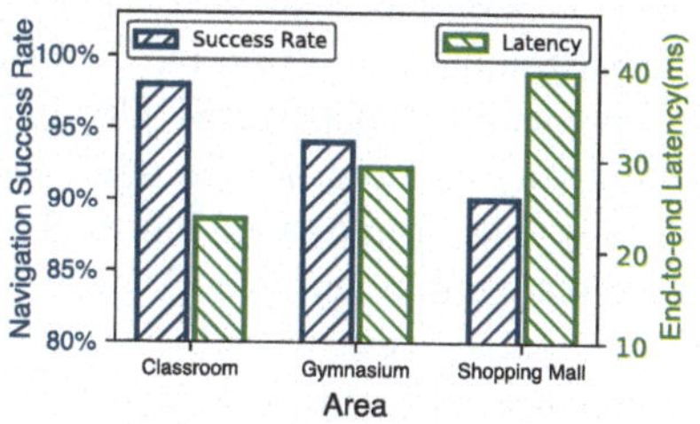

Fig. 3.17 Robot navigation success rate

similar performance (accuracy difference <20%) regardless of the environmental difference because of semantic mask transfer strategy.

3.6.6.2 Robot Mapping and Navigation

We conducted extensive experiments in an office building, a gymnasium and the first–third floor of a shopping mall, with area sizes of about 400 m^2, 1,000 m^2, and 4,000 m^2, respectively. Overall, we design 17 navigation paths, including four short paths ($\leq$100 m), six medium paths (100−200 m), and seven long paths ($\geq$200 m), covering all the main pathways of the testing areas.

Setup In this test scenario, client side of edgeSLAM is a robot implemented with an Nvidia Jetson TX2. The edge server has been mentioned above.

Evaluation Metrics Similar to some existing works like *Travi-Navi* and *Pair-Navi*, we set checkpoints at turns, escalators, and some landmarks on each trajectory. In total, we set 274 checkpoints for the 21 navigation paths. *Navigation success rate* is defined as the rate of successful arrival at each checkpoint.

Navigation Performance The performance of edgeSLAM is depicted in Fig. 3.17. The average navigation success rates (with NRCC) by edgeSLAM in classroom building, gymnasium, and shopping mall are 97%, 94%, and 90%, respectively.

Additionally, the wireless connections are different in three areas. In classroom building, client are connected to edge server under Wi-Fi-5G link, while in gymnasium and shopping mall are Wi-Fi-2.4G and Cellular-4G, respectively. The average end-to-end latency for each frame is also shown in Fig. 3.17. The result demonstrates that in classroom and gymnasium, edgeSLAM can run in real-time

(>30FPS) meanwhile about 25FPS in shopping mall. The rationale behind that is the network suffering severe fluctuation in crowded shopping mall.

3.7 Related Work

Visual SLAM Evolution The domain of Simultaneous Localization And Mapping (SLAM) has witnessed remarkable advancements over decades, facilitating its integration into extensive real-world applications. The initial breakthrough in monocular visual SLAM [30] employed a filter-based strategy, subsequently giving way to more precise optimization-driven methods [31, 32] as highlighted by [33]. The forefront of innovation, *ORB-SLAM* [7], utilized *DBoW2* [34] for place recognition and *g2o* [35] for optimization, marking a significant milestone. Contemporary research [36, 37] aims to weave semantics with visual SLAM to enhance adaptability in dynamic environments, underscoring its potential in powering robotic applications.

Advances in Semantic Segmentation The era of Convolutional Neural Networks (CNN) has ushered unparalleled success across various detection challenges, overshadowing traditional methods. Leading frameworks like Faster R-CNN [38], R-FCN [39], YOLO [40], and SSD [41] have set benchmarks in object detection and categorization. Recent innovations strive for segmentation at both semantic and instance levels. Initial attempts [42] combined R-CNN [43] for proposal classification with category-driven mask predictions. FCIS [44] integrated segmentation and detection, utilizing shared maps to capitalize on their interconnectedness. The pinnacle of this segment, Mask R-CNN [13], builds upon Faster R-CNN, introducing a concurrent branch for mask prediction alongside bounding box detection.

Edge-Assisted Innovations Leveraging edge or cloud infrastructures to handle computationally demanding tasks paves the way for seamless vision analytics and applications like VR/MR. Chen et al. [45] analyzed latency across several edge computing implementations. DeepDecision [20] formulated a strategy for dynamic task offloading based on network conditions. Solutions like VideoStrom [46] and Chameleon [47] optimize video analytics accuracy without increasing cloud computational demands. Liu et al. [17] introduced an edge-enhanced system ensuring precise object detection in AR/MR environments at 60FPS. While prior studies mainly address object detection, our focus on semantic visual SLAM presents a novel challenge, with segmentation and localization tasks proving significantly more intricate on mobile platforms.

3.8 Conclusion

In this chapter, we present the design and implementation of edgeSLAM, an edge-assisted visual SLAM system for mobile devices, achieving both accuracy and real-time simultaneously. The core technology of our design lies in: 1, the decomposition of computation modules of SLAM and semantic segmentation; 2, avoidance of redundant computation by reuse intermediate results; 3, adaption of system parameters to various conditions of network bandwidth and latency. We fully implement edgeSLAM and extensively evaluate the performance on three datasets, under different network conditions. The results show that edgeSLAM achieves delightful results in all scenarios. Being truly real-time, edgeSLAM sheds lights on practical localization and mapping for mobile users and robots.

References

1. C. Wu, J. Xu, Z. Yang, N.D. Lane, Z. Yin, Gain without pain: Accurate WiFi-based localization with fingerprint spatial gradient, in *PACM on Interactive, Mobile, Wearable and Ubiquitous Technologies* (2017)
2. J. Wang, D. Katabi, Dude, where's my card? RFID positioning that works with multipath and non-line of sight, in *Proceedings of the ACM SIGCOMM* (2013)
3. H. Liu, Y. Gan, J. Yang et al., Push the limit of WiFi based localization for smartphones, in *Proceedings of the ACM MobiCom* (2012)
4. J. Xu, Z. Yang, H. Chen et al., Embracing spatial awareness for reliable wifi-based indoor location systems, in *Proceedings of the IEEE MASS* (2018)
5. L. Yang, Y. Chen, X.-Y. Li, C. Xiao, M. Li, Y. Liu, Tagoram: Real-time tracking of mobile rfid tags to high precision using cots devices, in *Proceedings of the ACM MobiCom* (2014)
6. L. Shangguan, Z. Yang, A.X. Liu, Z. Zhou, Y. Liu, STPP: spatial-temporal phase profiling-based method for relative rfid tag localization. IEEE/ACM Trans. Netw. **25**(1), 596–609 (2016)
7. R. Mur-Artal, J.M.M. Montiel, J.D. Tardós, ORB-SLAM: a versatile and accurate monocular SLAM system. IEEE Trans. Rob. **31**(5), 1147–1163 (2015)
8. H. Abdelnasser, R. Mohamed, A. Elgohary et al., Semanticslam: using environment landmarks for unsupervised indoor localization. IEEE Trans. Mob. Comput. **15**(7), 1770–1782 (2016)
9. R. Mur-Artal, J.D. Tardós, ORB-SLAM2: an open-source SLAM system for monocular, stereo and RGB-D cameras. IEEE Trans. Rob. **33**(5), 1255–1262 (2017)
10. C. Cadena, L. Carlone, H. Carrillo et al., Past, present, and future of simultaneous localization and mapping: toward the robust-perception age. IEEE Trans. Rob. **32**(6), 1309–1332 (2016)
11. J.G. Andrews, S. Buzzi, W. Choi et al., What will 5g be? IEEE J. Sel. Areas Commun. **32**(6), 1065–1082 (2014)
12. I. C. S. L. S. Committee et al., IEEE standard for information technology-telecommunications and information exchange between systems-local and metropolitan area networks-specific requirements part 11: Wireless lan medium access control (mac) and physical layer (phy) specifications. IEEE Std 802.11^ (2007)
13. K. He, G. Gkioxari, P. Dollár, R. Girshick, Mask r-cnn, in *Proceedings of the IEEE ICCV* (2017)
14. J. Sturm, N. Engelhard, F. Endres, W. Burgard, D. Cremers, A benchmark for the evaluation of rgb-d slam systems, in *Proceedings of the IEEE IROS* (2012)
15. A. Geiger, P. Lenz, R. Urtasun, Are we ready for autonomous driving? The kitti vision benchmark suite, in *Proceedings of the IEEE CVPR* (2012)

16. R. Hartley, A. Zisserman, *Multiple View Geometry in Computer Vision* (Cambridge university press, Cambridge, 2003)
17. L. Liu, H. Li, M. Gruteser, Edge assisted real-time object detection for mobile augmented reality, in *Proceedings of the ACM MobiCom* (2019)
18. E. Dong, J. Xu, C. Wu, Z. Yang, Y. Liu, Pair-navi: Peer to peer indoor navigation with mobile visual slam, in *Proceedings of the IEEE INFOCOM* (2019)
19. *Object keypoint similarity (oks)* (2017). http://cocodataset.org/keypoints-eval
20. X. Ran, H. Chen, X. Zhu, Z. Liu, J. Chen, Deepdecision: A mobile deep learning framework for edge video analytics, in *Proceedings of the IEEE INFOCOM* (2018)
21. G. Bradski, The OpenCV Library, *Dr. Dobb's Journal of Software Tools* (2000)
22. J. Sanders, E. Kandrot, *CUDA by Example: An Introduction to General-Purpose GPU Programming* (Addison-Wesley Professional, Boston, 2010)
23. *Orb-slam2* (2016). https://github.com/raulmur/ORB_SLAM2
24. T.-Y. Lin, M. Maire, S. Belongie et al., Microsoft coco: Common objects in context, in *European Conference on Computer Vision* (Springer, Berlin, 2014)
25. *Pytorch mask r-cnn* (2017). https://github.com/multimodallearning/pytorch-mask-rcnn
26. *Grpc* (2017). https://grpc.io/
27. *Boost serialization* (2015). https://www.boost.org/doc/libs/1_70_0/libs/serialization/doc/index.html
28. *Protocol buffers* (2015). https://developers.google.com/protocol-buffers/
29. I.A. Bârsan, P. Liu, M. Pollefeys, A. Geiger, Robust dense mapping for large-scale dynamic environments, in *Proceedings of the IEEE ICRA* (2018)
30. A.J. Davison, Real-time simultaneous localisation and mapping with a single camera, in *Proceedings of the IEEE ICCV* (2003)
31. E. Mouragnon, M. Lhuillier, M. Dhome, F. Dekeyser, P. Sayd, Real time localization and 3d reconstruction, in *Proceedings of the IEEE CVPR* (2006)
32. G. Klein, D. Murray, Parallel tracking and mapping for small ar workspaces, in *Proceedings of the IEEE ISMAR* (2007)
33. H. Strasdat, J.M. Montiel, A.J. Davison, Visual slam: Why filter? Image Vision Comput. **30**(2), 65–77 (2012)
34. D. Gálvez-López, J.D. Tardos, Bags of binary words for fast place recognition in image sequences. IEEE Trans. Rob. **28**(5), 1188–1197 (2012)
35. R. Kümmerle, G. Grisetti, H. Strasdat, K. Konolige, W. Burgard, G^2o: A general framework for graph optimization, in *Proceedings of the IEEE ICRA* (2011)
36. S.L. Bowman, N. Atanasov, K. Daniilidis, G.J. Pappas, Probabilistic data association for semantic slam, in *Proceedings of the IEEE ICRA* (2017)
37. J. Civera, D. Gálvez-López, L. Riazuelo, J.D. Tardós, J. Montiel, Towards semantic slam using a monocular camera, in *Proceedings of the IEEE IROS* (2011)
38. S. Ren, K. He, R. Girshick, J. Sun, Faster R-CNN: Towards real-time object detection with region proposal networks, in *Advances in Neural Information Processing Systems* (2015)
39. J. Dai, Y. Li, K. He, J. Sun, R-FCN: Object detection via region-based fully convolutional networks, in *Advances in Neural Information Processing Systems* (2016)
40. J. Redmon, S. Divvala, R. Girshick, A. Farhadi, You only look once: Unified, real-time object detection, in *Proceedings of the IEEE CVPR* (2016)
41. W. Liu, D. Anguelov, D. Erhan et al., Ssd: Single shot multibox detector, in *ECCV* (Springer, Berlin, 2016)
42. B. Hariharan, P. Arbeláez, R. Girshick, J. Malik, Simultaneous detection and segmentation, in *ECCV* (Springer, Berlin, 2014)
43. R. Girshick, J. Donahue, T. Darrell, J. Malik, Rich feature hierarchies for accurate object detection and semantic segmentation, in *Proceedings of the IEEE CVPR* (2014)
44. Y. Li, H. Qi, J. Dai, X. Ji, Y. Wei, Fully convolutional instance-aware semantic segmentation, in *Proceedings of the IEEE CVPR* (2017)

45. Z. Chen, W. Hu, J. Wang et al., An empirical study of latency in an emerging class of edge computing applications for wearable cognitive assistance, in *Proceedings of the IEEE/ACM SEC* (2017)
46. H. Zhang, G. Ananthanarayanan, P. Bodik, M. Philipose, P. Bahl, M.J. Freedman, Live video analytics at scale with approximation and delay-tolerance, in *Proceedings of the USENIX NSDI* (2017)
47. J. Jiang, G. Ananthanarayanan, P. Bodik, S. Sen, I. Stoica, Chameleon: Scalable adaptation of video analytics, in *Proceedings of the ACM SIGCOMM* (2018)

Chapter 4
EdgeSLAM 2.0: Enhancing Scalability in Multi-Agent Systems

Abstract In the chapter, we delve into how SwarmMap (a.k.a., EdgeSLAM 2.0) revolutionizes collaborative visual SLAM services in edge computing environments, specifically tailored for burgeoning applications like search-and-rescue operations, inventory management automation, and industrial inspections. Central to SwarmMap's architecture are three innovative modules: a server–client synchronization process based on change logs, a scheduler attuned to task priority, and a streamlined global map representation. Together, they efficiently mitigate issues like data overflow, excessive bandwidth use, and inaccuracies in localization, which intensify with an increase in agent numbers. SwarmMap is made ROS-compatible, and its resources have been made publicly available (Access the code and resources at https://github.com/MobiSense/SwarmMap.). This enables existing visual SLAM solutions to readily adopt SwarmMap, significantly boosting their multi-agent handling capabilities. Through rigorous testing and a detailed 3-month deployment in one of the globe's most extensive oil-fields, SwarmMap demonstrated its ability to support double the agents ($\geq$20) compared to leading methods, while ensuring resource utilization remains constant. Additionally, it achieves a remarkable average trajectory accuracy of 38 cm, surpassing prior achievements by more than 55%.

4.1 Introduction

Visual SLAM technology, which processes video inputs from single or multiple cameras to simultaneously generate 3D environmental maps and pinpoint camera locations, has become a pivotal tool in navigation systems[1–3]. As environments for mapping grow in complexity, the adoption of multi-agent collaborative visual SLAM is seeing increased interest from research and commercial sectors alike [4–7]. Major e-commerce players like Amazon, JD, and Alibaba have implemented these systems in logistics centers with automated sorting robots to enhance efficiency [8]. Furthermore, aerial technology firms such as DJI and Amazon are advancing in drone formations for applications in city planning, expedited shipping, and field surveys [9]. These advancements underscore the necessity for agents to achieve real-time mapping and localization due to (i) the demand for updated maps

J. Xu et al., *Edge Assisted Mobile Visual SLAM*,
https://doi.org/10.1007/978-981-97-3573-0_4

by upper-tier applications for efficient task execution and management in dynamic settings and (ii) the intertwined nature of mapping and localization tasks, with each agent depending on a precise onboard map to improve localization accuracy and, conversely, mapping effectiveness [10, 11].

The SLAM agents profile the environment with their cameras, exchange data with each other, and execute vision tasks in real-time, with a significant computation overhead. The limited computation resource on the agent soon becomes the bottleneck, impairing system accuracy [4, 10, 11]. *Edge offload* has emerged as a promising alternative due to the following two reasons. First, by offloading bulky tasks to edge devices, the agents only need to run lightweight and time-sensitive jobs locally, which effectively mitigates onboard resource shortage [10, 11]. Second, by fusing and further optimizing the visual map globally at a centralized edge device, map information that is originally unavailable to each other can be easily shared among agents [4, 5]. This will benefit collaborative missions such as collision avoidance and path planning.

Albeit inspiring, the growing number of agents brings new issues that challenge the scalability of edge-based real-time collaborative visual SLAM systems (Sect. 4.2.2):

- **Map synchronization stresses the network bandwidth**. Mobile agents like drones and robots heavily rely on wireless links to communicate with an edge device. However, wireless spectrum is a limited and overcrowded resource. Streaming large volumes of map data over wireless links will soon saturate the medium and cause significant delays.
- **FCFS-based job scheduling impairs the localization accuracy**. An edge device has to processes large volumes of requests from agents, which may cause significant delays to latecomers (i.e., those requests positioned in the tail of the queue). However, agents in different states are not equally sensitive to the queuing delay. The conventional first-come first-served (FCFS) pipeline will exacerbate the localization error on those time-sensitive agents.
- **Map expansion exacerbates the memory footprint**. The size of the global visual map increases sharply with a growing number of agents, which is likely to exceed the limited memory capacity allocated to SLAM tasks by an edge node, causing memory overflow.

However, the current practice of edge offload focuses primarily on computation-oriented task partitioning [4, 10–13]. They fail to address the data explosion and its impact on transmission, scheduling, and storage. Hence these pioneer designs cannot scale with the sheer size of the real-time collaborative visual SLAM systems. In this chapter, we present SwarmMap, a framework to scale up the real-time collaborative visual SLAM services at resource-constrained edge devices. SwarmMap does not innovate visual SLAM algorithms. Instead, it proposes functionality and resource abstractions of existing SLAM algorithms and provides additional system services to enhance system scalability. Hence, most variations of collaborative visual SLAM systems can take advantage of our design. With SwarmMap, the upper-layer user can outsource agent task scheduling and processing instead of understanding every detail of SLAM algorithms to manually adapt. SwarmMap contains three key plug-in modules, as described below.

First, we design a Map Information Tracker (*Mapit*) to maintain map data consistency between the agents and the edge while remarkably saving network bandwidth. Unlike existing methods that transfer bulky map data with each other [4, 5], *Mapit* records the operations associated with the map modification on the agent and transmits these operations to the edge. The edge node then follows these operations to update its local map. This allows the map synchronization between them at the minimum bandwidth consumption even compared with state-of-the-arts (e.g., CarMap [14]).

Second, we introduce a SLAM-specific task-aware scheduler (*STS*) that prioritizes requests based on the status of their producer (i.e., agent). The *STS* scheduler runs on both the agent and the edge. The agent *STS* evaluates agent status around the clock and updates this information with the edge through heartbeat packets. The edge *STS* designs a multi-level queue to ensure those urgent tasks will be processed timely.

Third, we propose a Map Backbone Profiling (*MBP*) technique to alleviate the storage overhead while retaining the mapping accuracy. This technique is based on an observation that the data quality among different agents' maps can be balanced by elements in co-visible areas. We propose a set of metrics to detect high-quality map elements and use them to offset those low-quality counterparts, thereby elevating the overall map quality. Applying model compression to this high-quality map allows us to remove large portions of redundant map data without sacrificing the map accuracy.

We evaluate SwarmMap on a testbed consisting of four Nvidia Jetson boards, four smartphones, four DJI RoboMasters, and four drones. Following the standard SLAM evaluation pipeline [11, 14–16], we further compare SwarmMap with two state-of-the-art (SOTA) edge-assisted multi-agent SLAM systems (CCM-SLAM [4] and Multi-UAV [5]) on three gold-standard SLAM datasets (TUM [17], KITTI [18], and EuRoC [19]), as well as a self-labeled dataset collected at a 22,927 sqft shopping mall. We also compare SwarmMap with CarMap [14] and Sum-Map [20] to evaluate each functional module in SwarmMap. Our head-to-head comparison shows that SwarmMap can serve $2\times$ more agents than these SOTA systems with the same resource overhead, meanwhile maintaining an absolute trajectory error within 38 cm when serving 20 agents, outperforming these SOTA systems by >55%.

Real-World Deployment We have developed a real-time collaborative visual SLAM system based on SwarmMap and deployed it in one of the world's largest oil-fields (>170 km^2) for industrial inspection (shown in Fig. 4.1). Our system consists of 12 agents that communicate with an Nvidia Jetson AGX Xavier [21] edge node through Wi-Fi mesh networks. A 3-month pilot study shows that SwarmMap achieves an average localization accuracy of 0.36 m. The link throughput and RAM consumption are below 17 MB/s and 26 GB, respectively, meeting inspection demands within the constraints of available resources.

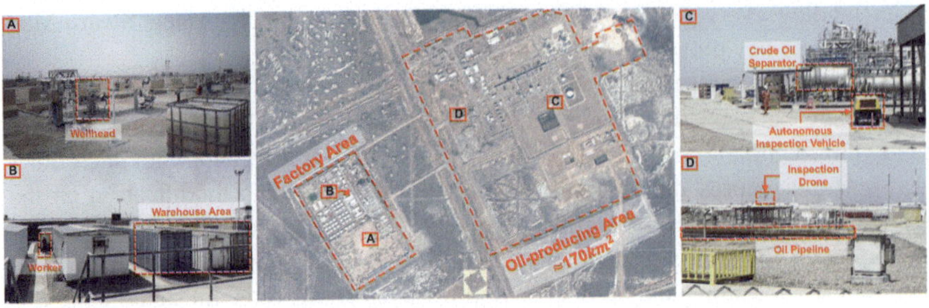

Fig. 4.1 Industrial inspection is carried out by ten drones and two autonomous vehicles in one of the world's largest oil-field (>170 km^2) in the Middle East. These agents are coordinated by SwarmMap that runs on an Nvidia AGX Xavier edge server

In summary, this chapter makes three contributions. First, we quantify the scalability challenges of deploying real-time collaborative visual SLAM at the edge to motivate framework support. Second, we design and implement SwarmMap as a framework to address the scalability issues spanning from communication, computation, to storage. As far as we are aware of, SwarmMap is the first system solution to scale up the collaborative visual SLAM in edge settings. Third, we deploy SwarmMap in one of the world's largest oil-fields for industrial inspections in the Middle East. Our 3-month pilot study demonstrates that SwarmMap makes a great process toward fortifying multi-agent collaborative visual SLAM to a fully practical system for wide deployment.

Contribution to the Community We implement SwarmMap as a software package of the robot operating system (ROS [22]), the dominating OS in the robotics field. We believe SwarmMap can provide a collection of tools for both academia and industry and further enable fast prototyping of visual SLAM-based applications in multi-agent scenarios.

4.2 Background and Motivation

The data volume scales with the number of agents, and the need for framework support arises from the excessive bandwidth consumption and memory footprint caused by the data explosion. We discuss these in detail in this section.

4.2.1 Edge-Assisted Visual SLAM Systems

The visual SLAM consists of multiple sub-tasks with diverse workloads. Edge offload places those bulky tasks to an edge server, leaving an agent lightweight

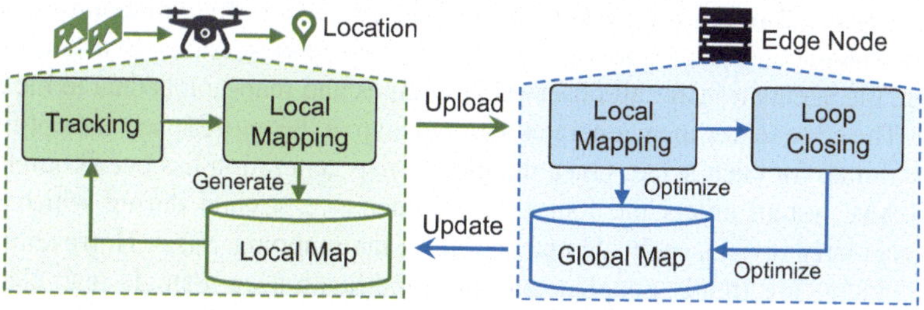

Fig. 4.2 Workflow of existing edge-assisted SLAM [10, 11]

and time-sensitive jobs. The agent can thus run visual SLAM in real-time. We use ORB-SLAM2 [1], a top-ranked open-source visual SLAM system, to illustrate the SLAM operations under edge settings (refer to Fig. 4.2).

Front-End Mobile agents run *Tracking* and part of the *Local Mapping* module locally. The *Tracking* module extracts 2D ORB feature points from each video frame and instantly estimates the pose of onboard camera(s) based on the geometry relationship between these feature points and the preconstructed local map (i.e., a set of 3D map-points and keyframes[1] in which they appear). As the mobile agent moves, the *Local Mapping* module updates the local map timely.

Back-End Due to high computation costs, the optimization part of the *Local Mapping* module is offloaded to the edge device, where the bundle adjustment (BA) algorithms [23] kick in to improve the pose and 3D location accuracy of those newly generated keyframes and map-points. The edge server also runs a *Loop Closing* module to detect repeated paths and leverage them to recalibrate the global map.

Data Transfer In-Between To improve the map accuracy, each agent periodically sends keyframes and map-points to the edge server for fine-grained optimization. The optimized visual map is then streamed to the clients.

4.2.2 *The Scalability Issues*

As more agents get involved, running real-time collaborative visual SLAM on edge environment becomes increasingly complex, facing several challenges: (i) the frequent data transfer between agents and edge is likely to saturate wireless links, causing significant delays, (ii) the queueing delay on edge node exacerbates localization errors, and (iii) the data volume grows sharply, threatening the data storage at the edge node. We discuss these issues below.

[1] Keyframes are a subset of selected frames. Each keyframe stores the camera pose, the map-points it observed, and the co-visibility relationships with other keyframes.

C1: Excessive Bandwidth Consumption The life cycle of a collaborative visual SLAM system consists of *cold-start* and *maintenance* two sessions. In the cold-start session, the agents transfer all observed keyframes and map-points data to the edge server. The edge server then generates a global map of the entire space and optimizes the local map for each agent. Once the global map generation has been completed, the SLAM system enters the long-term maintenance session during which each agent regularly revisits each site and calibrates the mapping offset. However, since map elements are tightly coupled, a minor modification on a single map element will spread to many other elements. This will cause a significant amount of data transfer in the maintenance.

To reduce bandwidth consumption, recent works [4, 14] design compact map representations and transfer the difference before and after map element calibration (as opposed to transferring the entire calibrated map element [5, 11]). Although these systems can effectively reduce bandwidth consumption in the cold-start session, they encounter two issues in the maintenance session due to the frequent map updates: (*i*) *extra computation overhead*. The acquisition of element-level differences requires pair-wise map feature comparison across the entire map. This will lead to extra computation workload pressure on resource-limited mobile agents. (*ii*) *Limited data volume reduction*. Since a minor change on an element will spread to a batch of coupled elements, the volume of data to be transferred is still bulky.

To validate our analysis, we measure the bandwidth requirement of three state-of-the-art (SOTA) systems in a different number of agent settings. The results are shown in Fig. 4.3a. Compared with the vanilla Multi-UAV [5], we observe that CarMap [14] and CCM-SLAM [4] can effectively reduce the transmission workload during map synchronization. However, when serving more than ten agents, both systems still produce excessive wireless traffic that can easily go beyond the link capacity;[2] thus, significant system delays are expected.

C2: Severe Localization Errors Under the edge settings, the localization accuracy of an agent highly depends on the quality of the local map which is optimized at the edge side. Typically, an agent needs to periodically (within 5 s) send optimization requests to the edge server for every 3–5 newly generated keyframes [10, 11]. As the number of agent scales, the concurrent requests from different agents block at the edge node's processing pipeline, resulting in excessive queuing delays. Consequently, some agents get their optimization tasks done untimely, causing severe localization errors. This situation is worsened by the fact that agents in different running states (e.g., flying speeds, self-tracking qualities) are not equally sensitive to the waiting delay. Recent multi-agent collaborative SLAM solutions focus on map fusion and optimization on edge or cloud servers but ignore the task queuing issue for each agent. The conventional first-come first-served (FCFS) scheduling will inevitably exacerbate the localization error on those task-sensitive agents (demonstrated in Sect. 4.5.3).

[2] The measurement shows the maximum throughput in an outdoor mesh and an indoor 2.4 GHz Wi-Fi network is 15 MB/s and 30 MB/s, respectively.

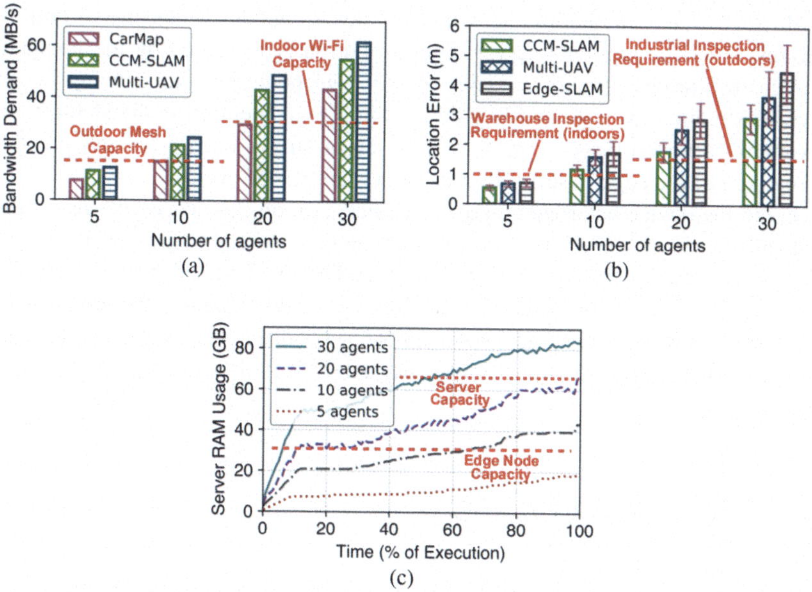

Fig. 4.3 The scalability issues in multi-agent scenarios. (**a**) Excessive bandwidth consumption. (**b**) Severe localization errors. (**c**) Large data storage overhead

We measure the localization error (in *m*) of three related works in a different number of agent settings. The results are shown in Fig. 4.3b. Considering the accuracy requirement from a broad range of SLAM applications, we treat 1 m and 1.5 m as acceptable localization errors for indoor (warehouse inspection) and outdoor (anomaly detection) scenarios. Evidently, all these three systems fail to meet the localization requirement when serving more than five and ten agent indoors and outdoors respectively, leaving room for improvements.

C3: Large Data Storage Overhead The global map maintained by the edge server contains large redundancy due to the following two reasons. First, to ensure the inspection efficacy, different agents will revisit the same area at certain intervals, causing significant path duplication. Second, to complete the 3D map reconstruction, different agents have to share a co-visible area, resulting in bulky data redundancy. As the number of agents grows, the data redundancy increases sharply, and the data volume is likely to exceed the limited memory capacity of the edge node.

We set up an edge-based collaborative visual SLAM testbed using a commercial edge device Nvidia Jetson AGX Xavier (with 32GB RAM and costs $599) and measure its RAM usage in different numbers of agent settings. We repeat the measurement on a powerful server with 4× higher storage capacity (i.e., Dell PowerEdge T630 with 128 GB RAM and costs $6,899) for comparison. The results are shown in Fig. 4.3c. In accordance with our analysis, as the system proceeds, the

RAM usage increases rapidly and soon saturates the memory capacity of both the edge node and the high-end server. This limitation is worsened by the mismatch between the limited storage capacity of the edge node and the growing fidelity of video streams (i.e., 4K or 8K videos). Such high memory demand limits the maximum number of agents to 5, which sets a strong barrier for the practical deployment of the edge-based collaborative visual SLAM system.

Due to the device heterogeneity (e.g., cameras on drones and robots may differ drastically in video resolution and frame rate) and diverse running status, the quality of maps provided by different agents may vary largely. An ideal map compression should remove those low-quality redundancies while retaining the high-quality counterpart. However, existing works ignore such difference when compressing the map data [20, 24, 25], resulting in degraded SLAM performance (details in Sect. 4.5.3).

4.2.3 SwarmMap: System Goals

SwarmMap takes a solid step forward in solving these scalability issues. We list the system goals below.

Goal 1: Functionality and Resource Abstraction SwarmMap should provide functionality and resource abstractions of existing SLAM algorithms. This allows any variation of map-point- and keyframe-based collaborative SLAM algorithms to take advantage of SwarmMap.

Goal 2: Plug and Play SwarmMap should be implemented as a plug-in module, exposing well-defined APIs to end-users for adaption. This avoids the deeply embedded manual code changes that may again challenge the system's scalability.

Goal 3: Resource Overhead Reduction SwarmMap should effectively reduce the resource overhead spanning data storage, client-edge communication, and task scheduling while ensuring the precision and real-time performance.

4.3 System Design for Edge-Assisted Multi-agent SLAM

In this section, we first describe the high-level system architecture and then present each module design in SwarmMap.

4.3.1 System Overview

SwarmMap is a framework design to scale up collaborative visual SLAM service in edge offloading settings. To achieve this goal, we make the following layer-wise functionality and resource abstractions: (i) *agent layer*, where each agent localizes itself and builds surrounding local maps in real-time, (ii) *network layer*, which enables communications and data interactions between mobile and edge for map synchronization, and (iii) *edge layer*, which fuses, optimizes, and maintains the global map. This layer-wise abstraction provides a clear view of map data transfer, processing, and storage in SLAMs.

Key Functional Modules SwarmMap designs three plug-in modules to address the resource overhead and scheduling issues across these three layers.

- The *Mapit* (Map Information Tracker) module tracks system operations associated with map data calibration. It then transfers these operations to the peer(s) for map synchronization (Sect. 4.3.2).
- The *STS* (SLAM-specific Task Scheduling) module optimizes the batch request execution and manages the resource allocations among multiple agents (Sect. 4.3.3).
- The *MBP* (Map Backbone Profiling) module compresses the map data uploaded by individual agents while ensuring the overall mapping accuracy (Sect. 4.3.4).

SwarmMap Architecture Figure 4.4 shows the system architecture. SwarmMap shares similar edge-based architecture with previous works and provides extra system support on both the mobile agent and edge server side, as discussed below.

- On the mobile agent side, SwarmMap tracks the run-time status of each agent through a lightweight evaluation-based mechanism *STS* (mobile part). It then follows a dedicated information exchanging protocol *Mapit* to communicate and update map elements with the edge server.
- On the edge side, the edge node prioritizes the agents' requests by *STS* (edge part) based on their run-time status. It then takes into account the data quality of maps reported by individual agents and extracts a lean presentation of the overall map through a map backbone profiling (*MBP*) algorithm. Finally, the optimized and compressed map backbones will be sent to each mobile agent by *Mapit*.

4.3.2 Mapit: Map Information Tracker

The inevitable frequent map data synchronization between clients and edge consumes large bandwidth in both cold-start and maintenance sessions, circumscribing the system capacity (i.e., the number of supported agents). Recent works (e.g., CarMap [14] and CCM-SLAM [4]) propose a compact map representation that

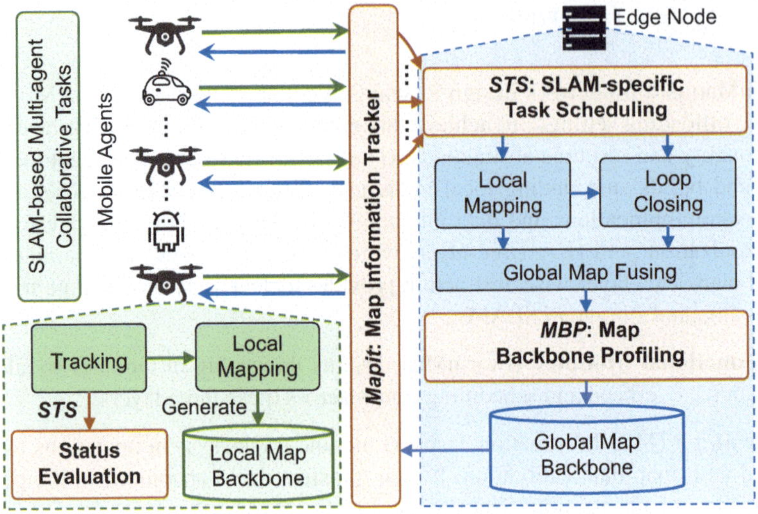

Fig. 4.4 System architecture of SwarmMap. Compared with the conventional edge-based visual SLAM architecture, the added plug-in modules are highlighted in orange

greatly reduces the data transfer in the cold-start session. However, their effectiveness fails to translate to a sufficient reduction in the maintenance session (Sect. 4.2.2-C1). Therefore, in SwarmMap we focus on the data transfer reduction in the maintenance session.

Our design is based on an observation that the map change on the one side can be reproduced on the other side (e.g., agent vs. edge) by solely transferring the map change operations. This enables a lightweight map synchronization by avoiding transferring massive map-point data and the bulky geographical descriptors such as their spatial locations, features, and observation relationships with keyframes [16]. Compared with the current practice, our design also achieves higher synchronization efficiency because it does not require a pair-wise map element comparison, which leads to extra computation workload pressure on resource-limited mobile agents.

To realize this basic idea, we design *Mapit*, a lightweight map information tracker to automate the operation tracking and reproducing on mobile and edge. *Mapit* runs as a daemon on both sides, monitoring the SLAM function calls and logging corresponding map operations (e.g., move a map-point by 2 cm). It then transfers this log to the agent (or the server), based on which the agent reproduces these operations locally. The map data are synchronized at the end.

The *Mapit* package periodically[3] synchronizes the map operation logs and consists of five atomic operations: *add*, *aggregate*, *push*, *merge*, and *pull* (shown in Fig. 4.5).

[3] Similar to current practice [5, 10, 14], we empirically set the period to 2 s.

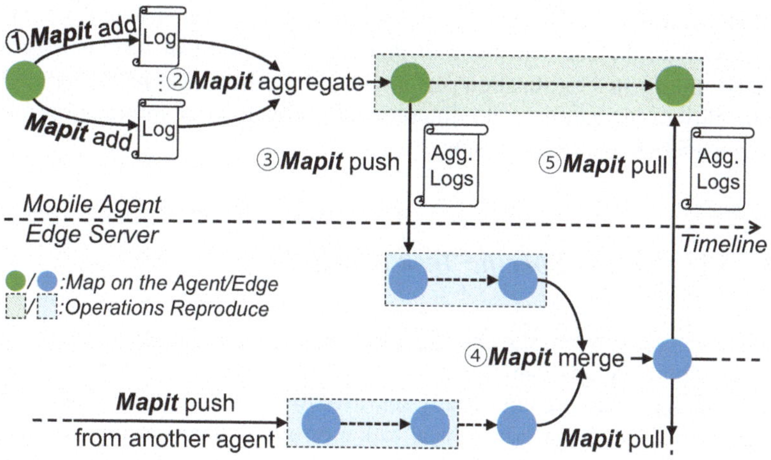

Fig. 4.5 Workflow of the *Mapit*

① ***Mapit* add**. The atomic operation *add* registers a hook for each SLAM function call (listed in Table 4.3) and maintains a recording queue. Whenever an important function is called, an operation record containing its name, parameters, and influence on map elements is *added* to the operation queue.

② ***Mapit* aggregate**. At the end of each period, *Mapit* aggregates the records in the operation queue to reduce their size. The intuition is that some removals or merges on certain types of functions will generate equivalent effects. For instance, if a function changes the location of a map-point and is marked as `overwritten`, we only need to focus on the latest record of it and ignore all the previous operations on the map-point. As for those marked as `stackable`, the implication is that records about modifying a same element can be merged by parameters. In this way, *Mapit* produces a minimal set containing necessary information.

③ ***Mapit* push**. After aggregating records, the atomic operation *push* on an agent sends the packed records to the edge server. By reproducing these operations, the map maintained on edge keeps synchronized with the ones on the client.

④ ***Mapit* merge**. On the edge server, the *merge* module periodically checks if there exists an overlap between the maps uploaded by individual agents and the global map. Once an overlap is detected, different maps will be coordinated and fused by the upper-layer SLAM algorithms (e.g., Sim3 optimization algorithm [16]). The map fusion process will operate and update some map elements, and hence the *merge* module also records these operations on the map elements in the same way as *add* and *aggregate*.

⑤ ***Mapit* pull**. The *pull* module can be treated as the reverse operation of *push*. It requests aggregated map modification logs generated by map optimization and *merge*, from the edge server to the agent. Additionally, if the global map has already been created (i.e., the whole system is in the maintenance session), *Mapit* will also transfer a set of closest map-points (e.g., associated with the next five keyframes) to the agent in the *pull* process. The benefit of this strategy is to enhance the agent's localization performance since these map-points with a high probability of appearing in the future would provide prior information for the *tracking* module on the agent side.

4.3.3 STS: SLAM-Specific Task Scheduling

As more agents get involved in SLAM systems, processing agents' requests (e.g., local map optimization) can cause excessive queuing delays. Since agents in different running states are not equally sensitive to the waiting delay, conventional FCFS scheduling may exacerbate localization errors on time-sensitive agents and hurt SLAM performance (Sect. 4.2.2-C2). To our best knowledge, there is still a lack of scheduling strategy tailor to multi-agent SLAM tasks.

To address this issue, we introduce *STS*—the first SLAM-Specific Task Scheduler that guides the edge to strategically prioritize requests. Specifically, *STS* divides agents into emergency and nonemergency groups based on the agents' status. It timely reorders the requests based on the following principles:

(i) Prioritizing requests from agents in the emergency group
(ii) Among those nonemergency agents, *STS* prioritizes requests from agents that can provide higher information gain for global map construction or optimization.

The first principle aims to prevent each agent from losing self-tracking, and the second is for achieving a better overall global mapping performance. We propose a set of metrics to characterize the agent status and design a multi-level queue to schedule the requests from agents.

4.3.3.1 Agent Status Evaluation and Updating

Agent Side Each agent regularly updates its status with the edge by sending heartbeat packets. Since both environment and device dynamics may fluctuate violently during an agent's movement, the heartbeat interval should be shorter than the agent's request interval (i.e., 2 s). In SwarmMap, we expose the heartbeat setting (100 ms by default) to end-users so that they can easily adapt to different environment settings. We define three variables that can fairly reflect an agent's status:

- *Tracking state*: a 1-bit Boolean value shows whether an agent is traceable or not. An agent's tracking state is set to LOST if its latest ORB feature maps cannot well match the local feature map. This variable is provided by the *tracking* module in many visual SLAM systems [1].

- *Velocity burst*: a 1-bit Boolean value shows whether an agent's speed changes abruptly or not. An abrupt change of velocity may result in motion blur in videos and make it hard for clients to extract visual features. In SwarmMap, we set the variable *Velocity burst* to True if the current moving speed is 20% greater than the average speed over the latest N frames, where N is a variable exposed to end-users. $N = 10$ by default.
- *Tracked map-point number*: an 8-bit variable represents the number of map-points observed by an agent. A larger number indicates the *tracking* module is running more stable.

Server Side Due to the heterogeneous device capability (e.g., cameras on different agents may differ in resolutions) and diversified trajectory, each agent contributes unequally to global map construction and optimization. SwarmMap prioritizes requests from those agents that can provide higher information gain for global map construction and optimization. To this end, we design the following two metrics to measure the information gain of each agent.

- *Map-point score* (MS) is defined as the average score of all map-points observed by an agent (the way to calculate the map-point score will be introduced in Sect. 4.3.4). A higher average score reflects that the current position is likely to have been visited before. On the contrary, a lower score indicates the agent is exploiting new or partially observed areas. Hence, *STS* prioritizes tasks with a lower map-point score.
- *Map element generation speed* (MG) characterizes the number of unobserved map-points and keyframes uploaded by the latest *Mapit push* operation. An agent with a higher map element generation speed contributes more to the edge's global map generation and optimization.

  STS normalizes each metric and computes each agent's *contribution score* as normalized MG—normalized MS.

4.3.3.2 Multi-Level Queue Scheduling

On the edge side, *STS* designs three queues with different priorities to facilitate agent request scheduling.

- **Lost Handling Queue**. If an agent's tracking state is marked as LOST, *STS* will push its request into this queue.
- **Lost Prevention Queue**. If an agent has a velocity burst and merely tracks few map-points, it may become prone to LOST, and *STS* will push its request into this queue.
- **Map Enrichment Queue**. For those agents with stable running status (i.e., without the risk of losing self-tracking), *STS* will push their requests into this queue and sort them by their mapping contribution scores.

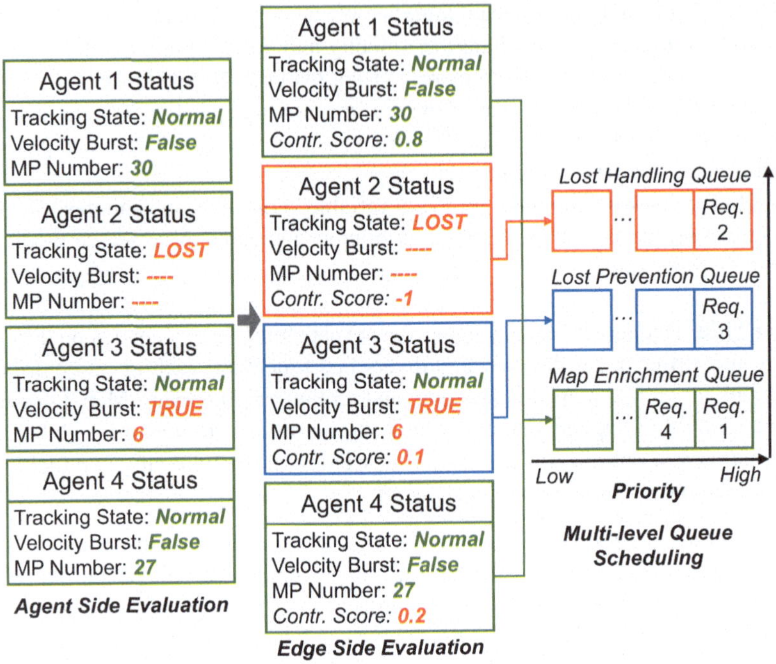

Fig. 4.6 Workflow of the *STS* with an example

The lost handling queue owns the highest priority, followed by lost prevention queue and map enrichment queue. Upon the reception of an agent's request, *STS* inserts this request into one of these three queues based on the agent's tracking status and mapping contribution. The back-end SLAM algorithm pops requests from queues based on their priority.

We take Fig. 4.6 as an example to explain the job scheduling in SwarmMap. Suppose there are four agents in the system, with agent 2 in lost tracking status and agent 3 facing the velocity burst issue. *STS* will push agent 2 and 3's requests into the lost handling and prevention queue, respectively. The request from agents 1 and 4, two agents not in emergency states, will be pushed into the map enrichment queue. Since agent 1's mapping contribution score is higher than agent 4, the request from agent 1 will be put at the head of the queue. The edge processes these requests in the order of 2–3–1–4.

4.3.4 MBP: Map Backbone Profiling

The global map maintained by the edge node contains large redundancy (Sect. 4.2.2-C3). Due to the device heterogeneity (e.g., the onboard cameras may differ in resolution and frame rate) and diverse running status, the quality of maps contributed by different agents may vary largely. Existing map compression works [15, 26, 27] ignore such difference, resulting in information loss and hence degraded performance. The relevant works, CarMap and CCM-SLAM, design lean map representations to reduce the transmitted data volume for a faster map synchronization. However, they still need to reconstruct the huge global map through these compact representations on both mobile agent and edge node. Therefore, the memory footprint remains high when more agents are connected.

To address this issue, we introduce a map backbone profiling (*MBP*) algorithm. Unlike the current practice, we do not greedily remove redundant map elements in co-visible areas. Instead, we first leverage these redundant elements to generate a series of virtual keyframes and use them to improve those low-quality map segments. Once the overall quality of the global map got improved, we can thus compress the global map without compromising the mapping quality.

MBP first evaluates the quality and importance of each map element. It then (i) finds high-quality map-points that could be leveraged to generate virtual keyframes, (ii) searches for low-quality map segments that need to be improved, and (iii) improves the overall map quality by inserting virtual keyframes to those low-quality map segments. Finally, *MBP* operates map compression on the balanced global map.

4.3.4.1 Map Element Evaluation

Map-point evaluation has been extensively studied in related works [27]. The gold-standard metrics include the observing path length, maximum observing distance, maximum observing angle, and mean reprojection error. We borrow these metrics (detailed in Sect. 4.9.2) to evaluate a map-point and propose three new metrics to adapt to collaborative scenarios:

- **Observed number** represents the number of keyframes, in which the map-point is observed, across the entire global map. A higher score indicates multiple agents can observe a map-point over a long period.
- **Update frequency** is defined as the total number of times the map-point was modified or updated by all agents in the last round of *Mapit push* operations. Map-points with high update frequency suggest a potential hot spot in a trajectory.
- **Moving velocity** records the speed of a mobile device when it generates the map element. A higher score indicates a potential blurriness that may influence the stability of the map-point. We take its negative value to evaluate the map-point score.

MBP normalizes each metric by its maximum value. We then define the score of a map-point as the sum of all normalized metrics values. The score of a keyframe is the sum of all observing map-point scores.

4.3.4.2 Map Backbone Generation

The map backbone generation consists of two steps: virtual keyframe generation and map compression.

Virtual Keyframe Generation The trajectory of an individual agent is first segmented with the awareness of where overlaps occur. The quality of each map segment is defined as the sum of all map element scores (i.e., scores of all keyframes and map-points) within it. For each map segment with low quality (e.g., its score is in the bottom 20%), *MBP* search for high-quality map-points in its neighborhood (i.e., within 60° field of view of its keyframes) even though the original keyframes do not observe these map-points. Furthermore, *MBP* synthesizes virtual keyframes that could observe these high-quality map-points, and the pose (i.e., spatial location and orientation) of each keyframe can be calculated by the ICP algorithm [28] and optimized by BA [23]. Since the virtual keyframes only consider whether a map-point is good enough regardless of which agent uploads it, they can supplement those low-quality segments.

Map Compression Once the quality of map segments is more balanced, *MBP* performs the similar map compression algorithm proposed by Sum-Map [20], eliminating redundancy by generating an enhanced minimum spanning tree across the global map. In addition, we introduce an extra optimization goal that guides the spanning tree to cover as many high-quality map elements as possible.

Figure 4.7 compares the map compression performance of *MBP* and Sum-Map. Map elements from different agents are marked in red, blue, and brown in the figure. Although Sum-Map obviously reduces the map size, it neglects the map quality difference, making the compressed map of trace 2 too sparse and harming the SLAM performance. In contrast, with reducing the map size by nearly half, *MBP* inserts several virtual keyframes, balancing the map quality among different agents and ensuring mapping accuracy.

4.4 System Implementation for Edge-Assisted Multi-Agent SLAM

We implement SwarmMap as an open-source package and make it compatible with ROS [22]. It contains 18,000 LOC (line of C++ code). SwarmMap is built upon ORB-SLAM2 [1], the top-ranked open-source SLAM algorithm that has been widely used by both research and industry communities. Our implementation avoids modifications on SLAM functions (e.g., tracking, local mapping, and loop closing). This allows any variation of ORB-SLAM algorithms such as DynaSLAM [29], ORB-SLAM3[30], and other map-point- and keyframe-based collaborative SLAM algorithms (e.g., Multi-UAV [5], C-ORB [31], and CCM-SLAM [4]) to take advantage of SwarmMap (demonstrated in Sect. 4.9.5). Additionally, we also

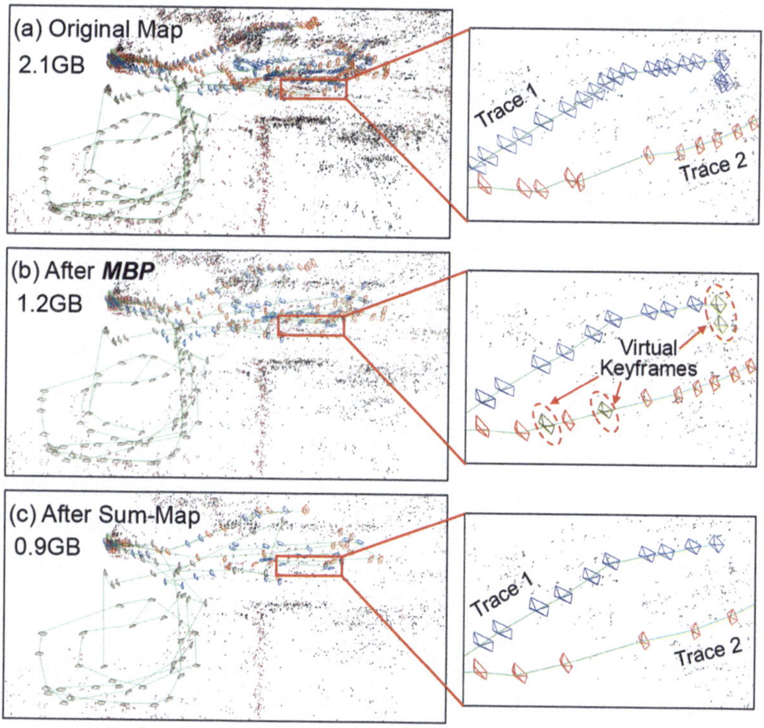

Fig. 4.7 A running demo of *MBP*. The left column shows the map elements uploaded by different agents, while the right column presents partial zooming-in maps. The dotted keyframes in (**b**) are the synthetic virtual keyframes

expose well-packaged APIs to facilitate users to modify some parameters (map synchronization period, status evaluation metrics, etc.) in SwarmMap according to specific upper-layer applications. A high-level abstraction of SwarmMap's implementation is detailed in Sect. 4.9.3.

4.5 Evaluation for Edge-Assisted Multi-Agent SLAM

In this section, we first present the experimental methodology (Sect. 4.5.1), followed by the overall performance of SwarmMap compared against SOTA systems (Sect. 4.5.2). We then conduct an ablation study to understand each functional module in SwarmMap (Sect. 4.5.3). Furthermore, we demonstrate the portability of SwarmMap by plugging it into baseline SLAM systems (Sect. 4.9.5).

4.5.1 Experimental Methodology

Field Studies We deploy 12 agents including 4 smartphones, 4 drones, and 4 mobile robots on a 22,927 sqft shopping mall. These agents collaboratively localize themselves and mapping the environment in real-time. The ground truth is obtained through the Kinect 360 RGB-D and Opti-Track [32] cameras. We also build a dataset using these video streams for trace-driven evaluation.

Trace-Driven Evaluations Following the conventional visual SLAM evaluation methodology [4, 11, 14, 31], we also conduct comprehensive trace-driven evaluations based on public SLAM datasets (KITTI [18], EuRoC [19], and TUM [17]) and the handcrafted dataset mentioned above. The characterization of three public datasets is summarized in Table 4.5. In our evaluations, the movement speed of mobile agents varies significantly, ranging from 0.5 m/s (indoor DJI RoboMasters) to 15 m/s (outdoor vehicles), representing the status of devices in real-world usage. Similar to the standard collaboration SLAM evaluation pipeline [4, 5, 33], we cut the video stream into overlapped segments and feed them to different agents to emulate the multi-agent scenario.

Edge Setup Most of the previous works cannot be deployed on a resource-constrained edge node to support large numbers of agents because they consume a considerable amount of network bandwidth and edge computational resources (Sect. 4.2). We thus use a powerful server, which is equipped with an Intel(R) Xeon(R) CPU E5-2620v4 of 2.10 GHz main frequency and 64 GB RAM running Ubuntu 18.04, to explore the capacity of these systems and compare them with SwarmMap. The agents communicate with the server through 2.4 and 5 GHz Wi-Fi links in the shopping mall and our laboratory. The maximally achievable link throughput measured with `iperf3` is 27.4 MB/s and 46.1 MB/s, respectively.

Metrics We use *absolute trajectory error* (ATE, in *cm*) to evaluate SLAM accuracy on the three public datasets while adopting *location error* (in *m*) to evaluate the positioning accuracy in field studies and our handcrafted shopping mall dataset. ATE is a golden metric for evaluating the tracking performance of SLAM algorithms [17]. Since ATE pre-calibrates the generated trace with the ground truth trajectories before measuring the absolute errors, it achieves fewer errors than the actual location errors. To evaluate system overhead, we count the *bandwidth demand* (in MB/s) of all participants in the system (defined as the sum of the average volume of data transferred per second by all agents). Similar to previous works [4, 11], we store the global map in RAM rather than SSD during system operation for faster map recall and update. We hence record the *RAM usage* (in GB) on the edge server to measure the memory consumption.

Map Updating Latency Similar to previous works such as Edge-SLAM [10] and CCM-SLAM [4], SwarmMap adopts the same edge-assisted architecture where the *tracking* task is running locally on the agents. This allows an agent to localize itself in real-time (i.e., >30 fps with camera rate). We thus take the *map updating* *latency* (in *ms*)—the delay until the agent gets the latest optimized map from the server—as the metric to evaluate the real-time performance of map updating in SwarmMap. Map updating latency takes into account both the map synchronization and optimization latency.

4.5.2 *Overall Performance Comparison*

We first compare SwarmMap with CCM-SLAM [4] and Multi-UAV [5], two most relevant SOTA edge-based multi-agent SLAM systems, to evaluate the overall performance.

4.5.2.1 Accuracy Comparison

We first evaluate the average ATE and location error in a different number of agent settings. The results are depicted in Fig. 4.8a and b. As seen, SwarmMap achieves the best tracking and localization performance in all scenarios. Compared with related works, SwarmMap reduces ATE by >30%, 20%, 20%, 50%, and 55% for scales with 3, 5, 10, 15, and 20 agents, respectively. The location errors are

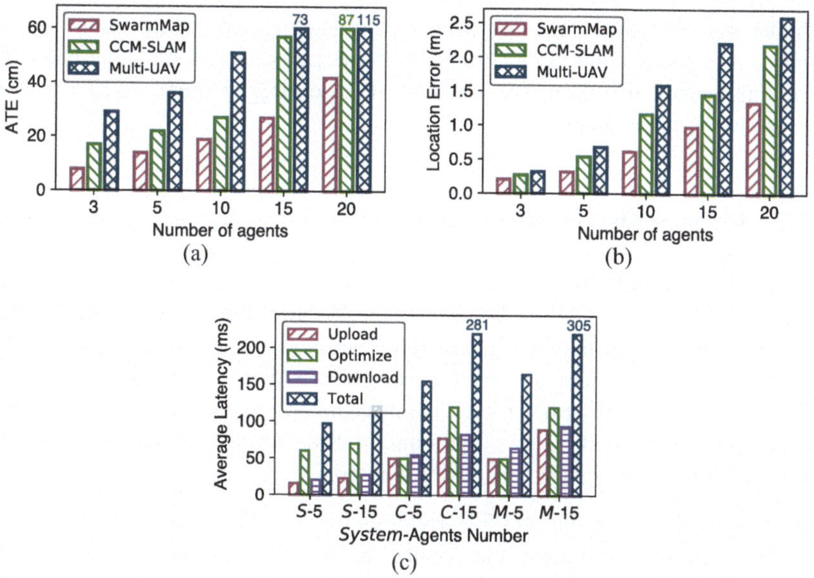

Fig. 4.8 Overall performance comparison with a growing number of agents. (**a**) ATE comparison on public datasets. (**b**) Localization error comparison on our shopping mall dataset. (**c**) Map updating latency comparison

also significantly degraded by >40% when serving more than ten agents. On the other hand, the performance of CCM-SLAM and Multi-UAV degrades remarkably with the growing number of agents (i.e., the ATE and location errors expand 3× and 7×, respectively, from 3 to 20 agents). When serving more than ten agents in the shopping mall, they fail to guarantee that the average location error of each client is within 1.5 m, which is typically the localization precision requirement for indoor drones [34]. In contrast, SwarmMap can still bound ATE and location error within 40 cm and 1.4 m even serving 20 agents. Generally speaking, above delightful results come from the fact that the localization performance of each agent highly depends on the quality of the onboard maintained local map [10, 11], and the three modules (*Mapit*, *STS*, and *MBP*) in SwarmMap exactly enable each agent to acquire an optimized local map in time.

4.5.2.2 Map Updating Latency Comparison

We further examine the end-to-end latency of each agent from uploading map segments to eventually obtaining the optimized map from the edge node. To save space in the figure, we denote SwarmMap, CCM-SLAM, and Multi-UAV as *S*, *C*, and *M*, respectively. Figure 4.8c shows the averaged latency on map uploading, optimizing, and downloading of each system in a different number of agent settings. As seen, the total latency of SwarmMap is around 95 ms and 105 ms for 5 and 15 agents respectively, outperforming baselines by >40% and 65%. The majority part of the latency reduction comes from the data uploading and downloading because *Mapit* reduces the amount of data transfers to a large extent. On the other hand, the processing latency also drops around 10% when serving 15 agents as *STS* module reduces the averaged queuing delay.

4.5.2.3 Resource Overhead Comparison

Bandwidth Demand We then measure the bandwidth demand of these three systems. As depicted in Fig. 4.9a, on average, SwarmMap reduces >35%, 20%, 30%, 25%, and 20% of network bandwidth requirement when serving 3, 5, 10, 15, and 20 mobile agents compared with existing works. Said differently, SwarmMap could serve more agents with the same wireless link throughput. For instance, under 27.4 MB/s shopping mall bandwidth limitation, SwarmMap can support more than 20 agents while existing works merely around 10.

RAM Usage We stitch the 00-05 trajectories on the KITTI dataset to generate a trajectory with 16.2 km length and conduct a 30 min experiment to measure the RAM usage. As shown in Fig. 4.9b, compared to CCM-SLAM, SwarmMap saves an average memory overhead of 2 and 6 GB when serving 5 and 15 agents, respectively, and the map size becomes stable under an upper bound (as seen, 15 GB when serving 15 agents) once the whole scene is well mapped. Unlike CCM-SLAM that

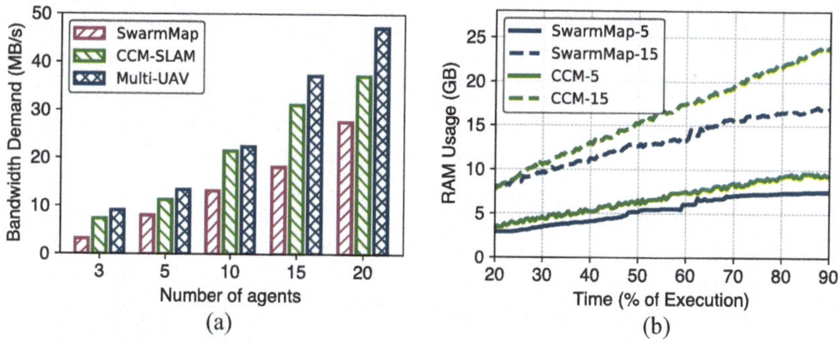

Fig. 4.9 Resource overhead comparison with a growing number of agents. (**a**) Bandwidth requirement. (**b**) Server RAM usage

requires transmission of a large volume of map elements, SwarmMap leverages *Mapit* and significantly reduces the bandwidth demand for map synchronization. In addition, the *MBP* module prunes the size of the global map maintained and optimized on the server, thus reducing the system overhead on computational resources.

Generally speaking, SwarmMap aims to scale the collaborative SLAM service with the same resource overhead, and will achieve a better performance with more computational resources are allocated on edge, which are left as future works.

4.5.3 Ablation Study

We then conduct an ablation study to understand the effectiveness of each module in SwarmMap.

Performance of *Mapit* We compare *Mapit* with CarMap [14], CCM-SLAM[4], and benchmark (e.g., edgeSLAM [11] and Edge-SLAM [10] that directly transmit the entire map without feature compression). Table 4.1 records the average data interaction speed (i.e., the average amount of map data uploaded and downloaded by each agent per second) of them on different datasets. As seen, *Mapit* saves nearly two times the bandwidth compared to CCM-SLAM and benchmark on all datasets. *Mapit* performs slightly worse than CarMap on KITTI and EuRoc datasets, where the operating environments are relatively large (e.g., broad city roads). In these scenarios, the agents spend most of their time in the cold-start session during which they continuously transfer the newly generated map elements. In contrast, on TUM and our shopping mall datasets, the SLAM system completes the environment profiling quickly and soon enters the maintenance session during which *Mapit* eliminates map data transfer and saves the bandwidth by adopting the strategy

Table 4.1 Transmitted data volume comparison

Solution	Average data interaction speed (MB/s)			
	TUM	KITTI	EuRoc	Shopping mall
Mapit	**1.3**	1.1	1.3	**1.4**
CarMap	1.9	**0.9**	**1.2**	1.8
CCM-SLAM	3.2	1.9	2.2	2.9
Benchmark	5.2	4.3	4.7	4.9

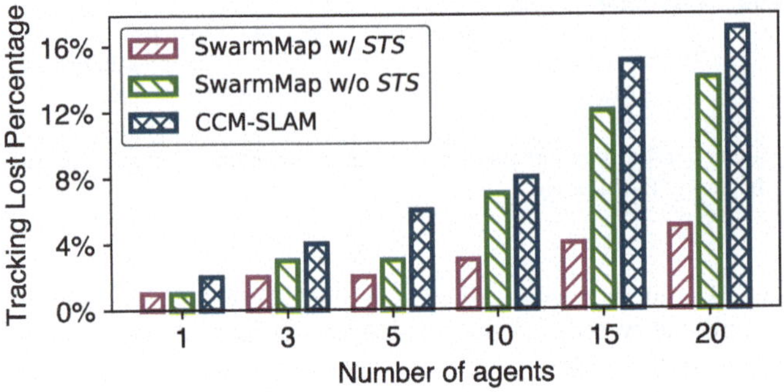

Fig. 4.10 Tracking stability comparison

of transmitting only records of map modifications rather than the modifications themselves.

Performance of *STS* We evaluate *STS* by counting the average tracking lost percentage (i.e., proportion of video frames, with which agents fail to track themselves, in all video frames) of SwarmMap with (w/) and without (w/o) *STS*. As depicted in Fig. 4.10, despite the increasing service scale, SwarmMap (w/ *STS*) maintains a stable service quality, and the lost percentage is within 4% in all scenarios. In contrast, the lost percentage of CCM-SLAM, as well as SwarmMap (w/o *STS*), increases rapidly, and the average lost percentage is at least 8% when serving more than ten agents, which may lead to a terrible self-tracking and environmental mapping performance. Generally speaking, the *STS* strategy enables SwarmMap to prioritize tasks depending on the agent emergence states and prevent most agents from losing self-tracking.

Performance of *MBP* We finally compare *MBP* with a map compression algorithm Sum-Map [20]. Specifically, we evaluate the map size after compression by their approaches and, equally important, the localization accuracy of each agent using the compressed map for self-tracking. The results are recorded in Table 4.2. We conduct experiments on the KITTI 02 and 05 trajectories because of the large map redundancy in them. The benchmark (only stores the global map without compressing it) shows the size of the original map and the ATE by using it. As seen, *MBP* reduces the original map size by almost half. Although the map compression ratio of *MBP* is a little smaller than that of Sum-Map, *MBP* barely sacrifices the accuracy of the global map.

Table 4.2 Map compression performance comparison

Solution	KITTI 02		KITTI 05	
	Map size (GB)	ATE (cm)	Map size (GB)	ATE (cm)
MBP	3.1	**7.6**	1.9	**6.4**
Sum-Map	2.8	10.7	1.8	9.3
Benchmark	5.2	7.4	4.1	5.8

4.6 Oil-Field Case Study

Based on SwarmMap, we have developed a real-time collaborative visual SLAM system and deployed it in one of the world's largest oil-fields ($>170\,\text{km}^2$) in the Middle East for industrial inspection. The details about the deployment setups can be found in Sect. 4.9.6. We conduct a 3-month pilot study (from June 2021 to August 2021) and summarize our main findings regarding the SLAM accuracy and system overhead.

SLAM Accuracy We calculate the average location error of each agent during inspections and present these results in Figs. 4.11 and 4.12. Note that we cannot directly obtain ground truth in the same way as in the experiment (e.g., deploy expensive Lidar or Opti-Track cameras); hence, we collect the video frames captured by all agents and run the multi-agent ORB-SLAM3 offline afterward without considering the system latency. On this basis, we take the difference between the real-time localization performance of SwarmMap and offline processed results as the location error. As shown in Fig. 4.11, the average location error is 19.3 cm and 29.1 cm in indoor and outdoor scenarios, respectively, satisfying the task requirement (1 and 1.5 m for indoor and outdoor inspections). Figure 4.12

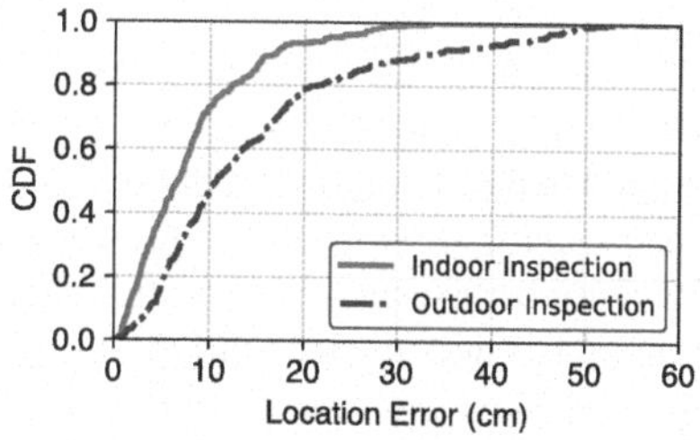

Fig. 4.11 Different scenarios accuracy

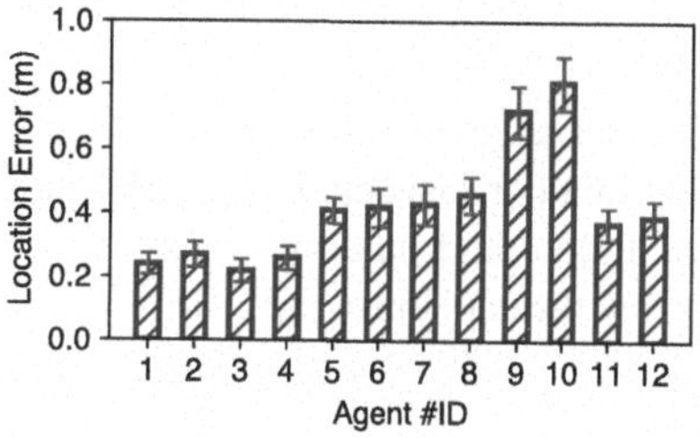

Fig. 4.12 Accuracy of different agents

Fig. 4.13 Latency measurement

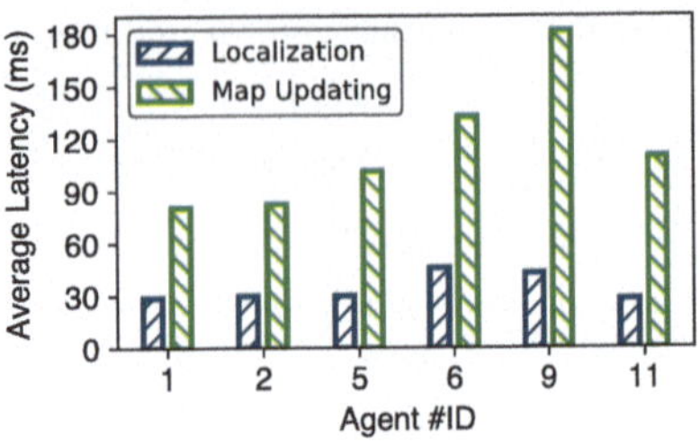

further illustrates the performance of each agent, and we find that two outdoor inspection drones (agents 9 and 10) suffer from a higher location error (up to 1 m). The reason behind it is that these two drones are carrying out oil pipeline inspection at the border of the oil field; they fly faster (e.g., >5 m/s) and far away from the edge server (e.g., 15 km). Therefore, they may experience certain delays due to the data forwarding through multi-hop mesh networks. Such a transmission delay may set a barrier for the drones to obtain the optimized map in time, causing localization errors. Nevertheless, the worst localization error of these two drones still satisfies the localization requirement in the outdoor scenario.

Latency We measure each agent's onboard localization latency (the delay on estimating its own location from an input image) and map updating latency. The results are depicted in Fig. 4.13. We observe that each agent could localize itself in a real-time manner (i.e., the localization delay is within 35 ms, typically the camera inter-frame interval). The average map updating delay is around 100 ms. Although agent 9 suffers from a higher map updating delay (an average of 191 ms) due to multi-hop data forwarding, it can still localize itself in real-time by leveraging its local map data.

Bandwidth Demand We record the total bandwidth demand for indoor (four agents) and outdoor (eight agents) inspection tasks. Figure 4.14 shows a snapshot over a span of 175 min. We find there is a drop in bandwidth demand at 45 min and 75 min, respectively. This is because the SLAM system enters the maintenance session at these two time points. Thanks to *Mapit*, the transferred data volume in the maintenance session is significantly reduced, with 4 MB/s for indoor and 11 MB/s for outdoor inspections. Additionally, due to the relatively higher flight speed and map updating delay for outdoor drones, the edge server needs to frequently transmit updated maps to them in *Mapit pull* to prevent them from losing self-tracking, which results in the outdoor bandwidth demand fluctuation more dramatically than indoor ones.

RAM Usage We further record the edge's RAM usage when executing the indoor and outdoor inspection tasks. As shown in Fig. 4.15, the maximum RAM usage in the indoor and outdoor scenarios is around 20 and 12 GB, both of which are well below the capability (32 GB) of the edge node.

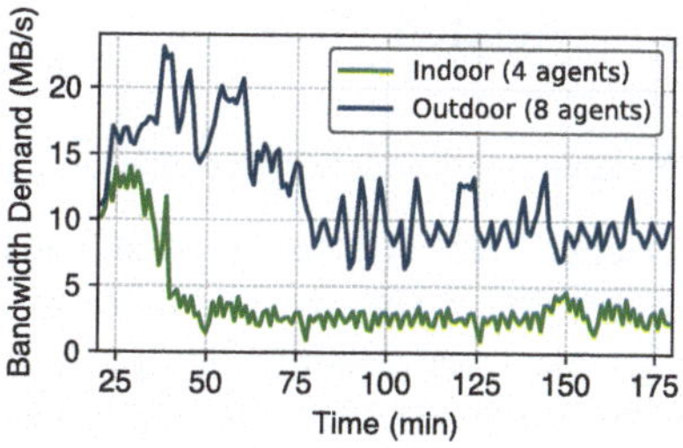

Fig. 4.14 Bandwidth consumption

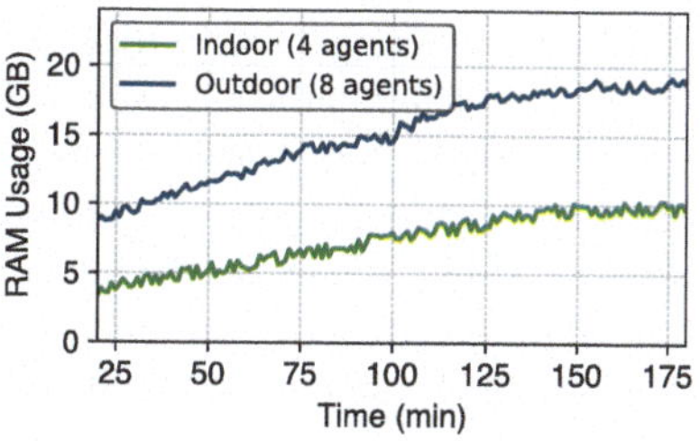

Fig. 4.15 Edge server RAM usage

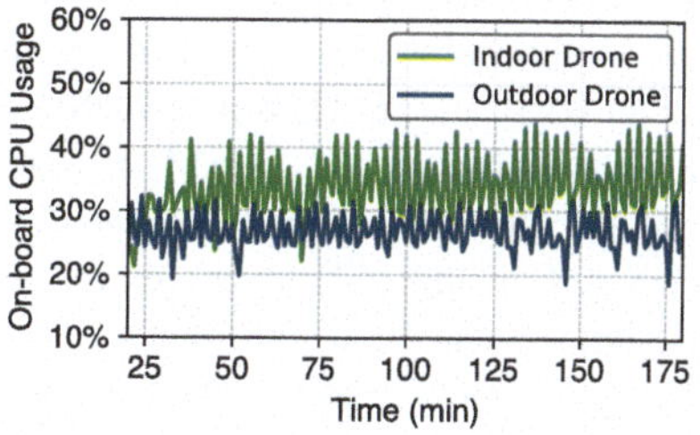

Fig. 4.16 Mobile CPU occupation

Onboard CPU Usage We also record the CPU occupancy rate of SwarmMap task (mobile part) on agents 1 (indoor drone) and 6 (outdoor drone) and plot these results in Fig. 4.16. The CPU usage of the outdoor drone is in the range of 20–35%, while the indoor drone is 22–43% during the 210 min of inspections. Due to the high dynamics of the indoor environment, the agent has to frequently update the local map although the whole area is well mapped, which takes up more CPU resources than outdoor environments. Note that SLAM is an underlying algorithm that provides an agent with location and environmental information, and SwarmMap still leaves more than 50% CPU computational resources for each agent to perform upper-layer applications (e.g., context-aware interaction, object detection, or segmentation).

4.7 Related Work

We review the most related works in this section.

Visual SLAM One of the most fundamental algorithms in robotics has been a topic of research in robotics and mobile systems for several decades [15]. It consists of the concurrent construction of a surrounding environment and the state estimation of the robot moving within it. Typically, systems use monocular cameras [35, 36], stereo cameras [1], or RGB-D cameras [37]. Some of the more well-known visual SLAM examples include RGBD-SLAM [38], RTAB-Map [39], and ORB-SLAM [1, 16, 30]. Although SwarmMap is implemented on the top of ORB-SLAM2 [1], it can be easily ported to other map-point-based visual SLAM like S-PTAM [2]. Other multi-map merging or optimization algorithms leveraged in recent work like ORB-SLAM3 [30] can also be integrated into SwarmMap. Our platform can also be applied to some feature/map-point-based multi-sensor SLAM systems like VI-ORB [40], VINS[41], and mmWave SLAM [42, 43].

Edge-Assisted Real-Time SLAM Recent studies [4, 10–14] speed up the computation-intensive tasks on agents by task partition and offloading workload to an edge server. Therein, edgeSLAM [11] and Edge-SLAM [10] enable mobile agents to run visual SLAM in real-time. They split the original ORB-SLAM2 architecture and offload the *local mapping* and *loop closure* tasks to an edge server. CarMap [14] leverages the map constructed by crowdsourced agents and designs a near real-time map update framework between client and cloud. Muti-UAV [5] and CCM-SLAM [4] leverage a central server with potentially larger computational capacity to merge and optimize maps constructed by different agents, while each agent maintains partial local maps for tracking. However, as the number of serving agents scales, these works face severe scalability issues including excessive bandwidth consumption, severe localization errors, and large data storage. SwarmMap is the first work that solves these scalability issues based on the same edge settings.

Multi-Agent Collaborative SLAM Collaborative SLAM has been explored recently [15]. C2TAM [44], C-ORB [31], and CVI-SLAM [33] present collaborative SLAM frameworks based on PTAM [35], ORB-SLAM2 [1], and VI-ORB [30], respectively. CSfM [45] also proposes a framework to coordinate maps upload from different agents. In general, the system goals of these works and SwarmMap are orthogonal: Above systems mainly focus on map fusion, optimization, and segmentation to generate a high-quality global map of the environment, ignoring the real-time performance of each agent and the entire system. In contrast, SwarmMap aims at solving the scalability issues and supports each agent for real-time tracking, mapping, and map updating. Inspired by current efforts, we could integrate some map merging, optimizing, and even compressing algorithms proposed by recent works [6, 15, 20, 26, 27] into SwarmMap for a better SLAM performance, which are left as future works.

4.8 Conclusion

We have presented the design and implementation SwarmMap, a framework to support real-time collaborative visual SLAM at edge devices. SwarmMap proposes functionality and resource abstractions of SLAM systems and provides three lightweight system services to address the communication, storage, and scheduling issues in edge-based scenarios. We implement SwarmMap as a software package on the ROS platform so that most variations of visual SLAM systems can directly benefit from it. Extensive evaluations and a 3-month pilot study demonstrate its superior performance.

4.9 Essential Supplementary Materials

4.9.1 *Functions Registered in* Mapit

In the *Mapit add* module, we dig the insights about how map elements get changed and find these changes mainly caused by certain important SLAM functions, a fraction of which is listed in Table 4.3. Thus, modifications that happened to the map can be recorded as calling history of these functions. For certain functions shown in the table, some removal and compression on the records will not harm data consistency. For instance, if a function is marked as `overwritten`, it indicates that its only effective change on a map element is the latest one i.e., changing the pose of a map-point. As for those marked `stackable`, the implication is that records about modifying the same element can be merged by parameters and still yield the same effect.

4.9.2 *Map-Point Evaluation Metrics*

A typical SLAM map consists of two types of elements, map-points and keyframes. Map-points represent discrete 3D landmarks in the global coordinate, and keyframes are selected frames indicating poses and positions of the corresponding camera (as illustrated in Fig. 4.17 with corresponding notations in Table 4.4). EBM [27] introduces several features based on local geometry information; we list four important metrics to evaluate a map-point we used in *MBP*:

- **Observing Path Length**. It is the distance traveled while observing the map-point and is obtained as

$$\phi_d^i = \sum_{j \in \mathcal{S}^i} \|\mathbf{t}_{\mathrm{G}}^{j+1} - \mathbf{t}_{\mathrm{G}}^j\|_2.$$

Table 4.3 Functions that could change the map element (only some fundamental functions are listed)

Target	Function	Type	Description
KeyFrame	`SetPose`	Overwritten	Set the pose of the keyframe
KeyFrame	`AddMapPoint`	Unique	Add a map-point to the keyframe
KeyFrame	`EraseMapPointMatch`	Unique	Remove a map-point from the keyframe
KeyFrame	`SetBadFlag`	Unique	Mark the keyframe bad and delete it
MapPoint	`SetWorldPos`	Overwritten	Set map-point-position in the world coordinate
MapPoint	`AddObservation`	Unique	Add a keyframe that observes the map-point
MapPoint	`EraseObservation`	Unique	Remove a keyframe from observations
MapPoint	`SetBadFlag`	Unique	Mark the map-point bad and delete it
MapPoint	`IncreaseVisible`	Stackable	Increase the count that map-point is observed
MapPoint	`IncreaseFound`	Stackable	Increase the count that map-point is matched
MapPoint	`SetLastTrackedTime`	Overwritten	Set the last tracked time of the map-point
MapPoint	`UpdateNormalAndDepth`	Overwritten	Update the normal vector and depth of the map-point
Map	`Clear`	Overwritten	Clear the current map
Map	`AddLoopClosing`	Unique	Add a keyframe to loop closing queue

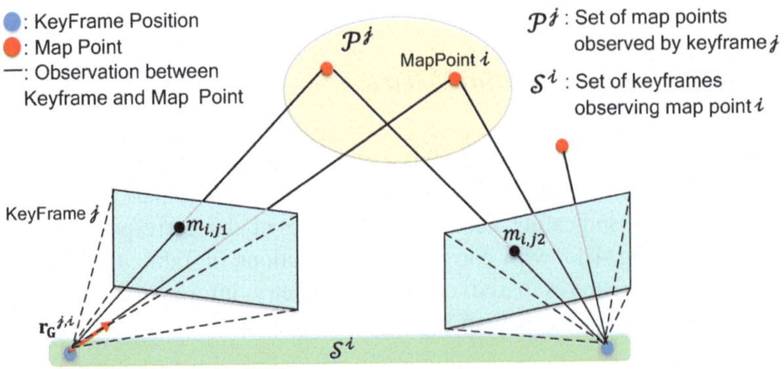

Fig. 4.17 Observation connection between keyframes and map-points

Table 4.4 Notation description

Notation	Description
$\mathbf{X}_{\mathrm{G}}^{i}$	Position of map-point i in global coordinate G
$\mathbf{t}_{\mathrm{G}}^{j}$	Position of keyframe j in global coordinate G
$\mathcal{S}^{i}$	Set of all keyframes observing map-point i
$\mathbf{r}_{\mathrm{G}}^{j,i}$	Unit-length observing vector starting from the observing keyframe j to map-point i in global coordinate G
$\mathcal{P}^{j}$	Set of all map-points observed by keyframe j
$\mathcal{M}^{i}$	Set of all agents observing map-point i
t_c^i, t_l^i	Creation and last tracked time for map-point i

- **Maximum Observing Distance**. It is the distance traveled between two most distant keyframes on a track, and each of them observes the map-point. Its computation requires maximization over all keyframes observing the same map-point, i.e.,

$$\phi_{\delta}^{i} = \max_{j,k \in \mathcal{S}^{i}} \|\mathbf{t}_{\mathrm{G}}^{j} - \mathbf{t}_{\mathrm{G}}^{k}\|_2.$$

- **Maximum Observing Angle**. It is the maximum angle between two keyframes that could observe the map-point and is obtained as

$$\phi_{a}^{i} = \max_{j,k \in \mathcal{S}^{i}} \arccos(\mathbf{r}_{\mathrm{G}}^{j,i} \cdot \mathbf{r}_{\mathrm{G}}^{k,i}).$$

- **Mean Reprojection Error**. Apart from the map-point track geometry, it is also worth considering the consistency of the map in the map-point's locality. EBM calculates the average reprojection error of each map-point to represent the mapping stability, i.e.,

$$\phi_{p}^{i} = \frac{\sum_{j \in \mathcal{S}^{i}} \|m_{i,j} - m'_{i,j}\|_2}{|\mathcal{S}^{i}|}.$$

4.9.3 SwarmMap Abstraction

Figure 4.18 shows the high-level abstraction of SwarmMap's implementation. The *MBP* module assists the map fusion and optimization unit to eliminate the data redundancy in the global map. The *STS* module replaces those handcrafted request handlers in conventional SLAM implementations [4, 5, 33] and thus alleviates the end-users' development overhead. Finally, we replace the communication unit

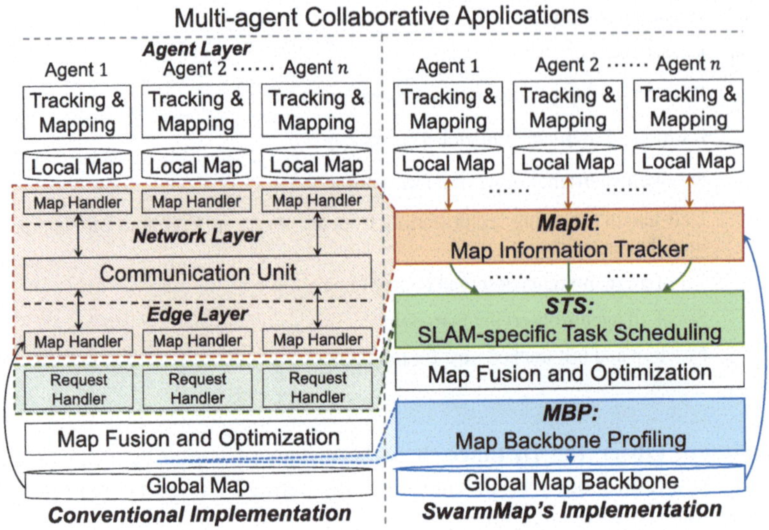

Fig. 4.18 High-level abstraction of SwarmMap's implementation. The arrow shows the data flow

and map handlers with a unified *Mapit* module. Such a layered implementation decouples SwarmMap's functional modules, allowing the end-users to turn on/off each module as they need. It also avoids the deeply embedded manual code changes (e.g., defining handlers) that again challenge the system scalability.

4.9.4 Experimental Dataset Description

We list the public datasets, the trajectories we used, and our handcrafted shopping mall dataset in Table 4.5. We select representative trajectories with different difficulty levels (in terms of environmental dynamics, path length, feature point sparsity, ambient light intensity, etc.) in TUM, KITTI, and EuRoc datasets, respectively.

4.9.5 Plug and Play

We demonstrate the portability of SwarmMap by integrating each of its components into two different SLAM systems. We add *STS*, *Mapit*, and *MBP* to ORB-SLAM3 [30], the latest follow-up of the ORB-SLAM system, and measure the accuracy gain brought by each module. Figure 4.19 shows the results. As seen, all these three modules contribute to localization accuracy. When serving five agents, *STS*, *Mapit*, and *MBP* decrease location errors by around 50%, 30%, and 10%, respectively. The contribution of each component also grows with an increasing number of agents.

Table 4.5 Dataset description

Dataset label	Trajectory sequence	Total time (min)	Total path (m)	Total frames	Environment
T-M (TUM medium and easy)	fr2_desk	1.66	18.88	2965	Office
	fr3_long_office_household	1.45	21.46	2585	
T-D (TUM difficult)	fr2_large_with_loop	2.88	39.11	5182	Industrial hall
	fr2_large_no_loop	1.87	10.93	3359	
K-M (KITTI medium and easy)	00/05	7.57/4.79	3724.18/2205.58	4541/2761	City road
K-D (KITTI difficult)	02/04	7.77	5067.23/393.65	4661/271	City road
E-M (EuRoC medium and easy)	MH_01/ MH_02	2.47/2.50	68.52/73.50	3682/3040	Machine hall
E-D (EuRoC difficult)	MH_04/MH_05	1.65/1.85	91.70/97.59	2033/2273	Machine hall
Shopping mall (our dataset)	N/A	15	314.2	24,365	Shopping mall

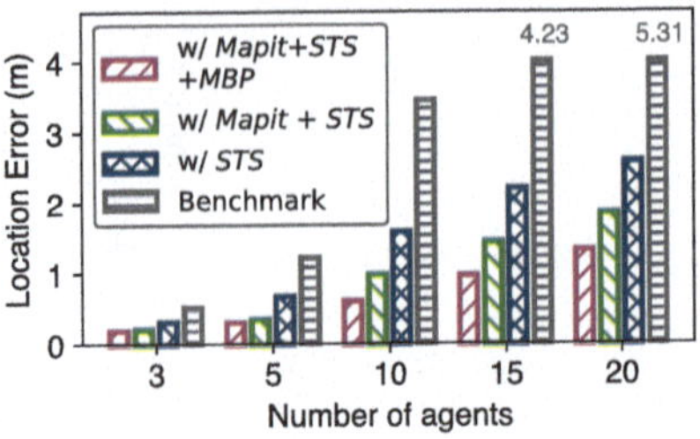

Fig. 4.19 Performance of each module

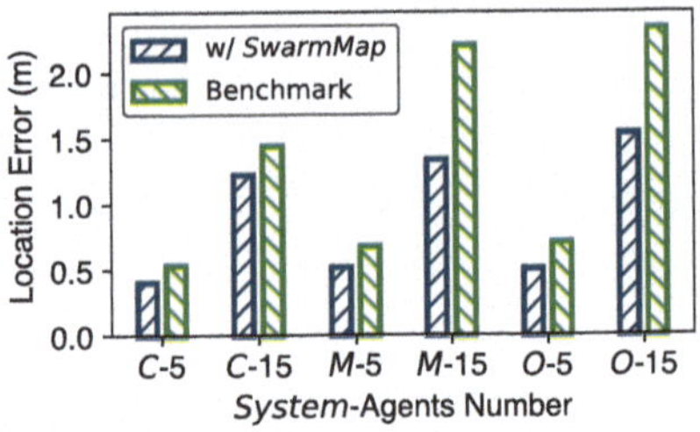

Fig. 4.20 Performance gains

We also integrate SwarmMap into our baselines CCM-SLAM, Multi-UAV, and ORB-SLAM3 (abbreviated as *C*, *M*, and *O*, respectively) to explore the location error reduction. As depicted in Fig. 4.20, the location error of CCM-SLAM, Multi-UAV, and ORB-SLAM3 decreases by 13.4%, 12.2%, and 16.7%, respectively, in five agents' settings. When serving 15 agents, the error decreases further to 17.2%, 31.3%, and 29.6%.

Remarks These results show that most existing works in multi-agent scenarios (especially scenarios with more agents) can directly benefit from SwarmMap. It is worth mentioning that we do not redesign or modify the code structure of these existing works for integration. We merely provide a wrapper to hook up these systems and SwarmMap (i.e., call the API defined in SwarmMap).

4.9.6 *Case Study Setups*

Our system consists of 12 mobile agents to perform daily inspection tasks both indoors and outdoors. These agents communicate with an Nvidia Jetson AGX Xavier edge node through Wi-Fi mesh networks, as shown in Fig. 4.21.

Inspection Agents We have deployed 12 mobile agents to perform daily inspection tasks, including four DJI Inspire drones (Agent #ID 1-4, equipped with 2K cameras) for indoor warehouse inspection, as well as six DJI Inspire (#ID 5-10) and two inspection vehicles (#ID 11-12, equipped with 1080P cameras) for outdoor oil-field inspection. For drones, we integrate the mobile part of SwarmMap into ArduPilot[46], a widely used open-source drone development platform. The output localization and mapping results are streamed to the Ardupilot Mega controller

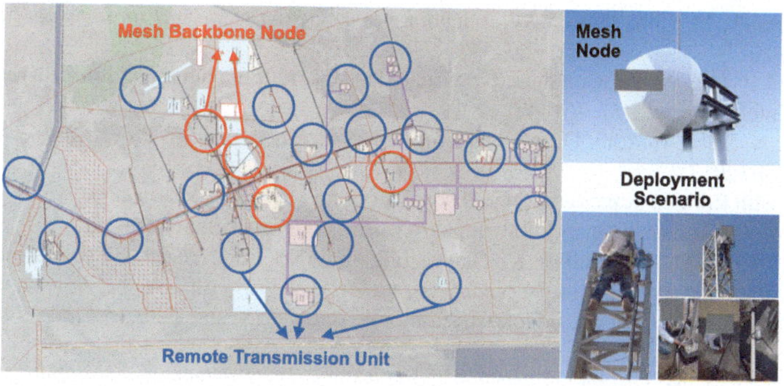

Fig. 4.21 Mesh network deployment in the oil-field

through a Micro-USB port for supporting upper-layer applications (e.g., real-time drone flight control, abnormal events detection). The two inspection vehicles are equipped with Nvidia Jetson TX1 as their computing units.

Edge Server We implement the edge side of SwarmMap on an Nvidia Jetson AGX Xavier edge node with a 32-GB 256-Bit LPDDR4x RAM, a 16-core ARM v8.2 64-bit CPU, and a 512-core Volta GPU. We also turn on the GPU acceleration by Numba [47] and CUDA [48] to speed up the back-end global map optimization procedure. The power consumption of the edge node is below 30 W, which is less than the available power supply in the industrial scenario.

Wireless Network The four indoor inspection drones communicate with the edge node via 2.4 GHz Wi-Fi, while the eight outdoor inspection agents communicate through a mesh network. In order to make the mesh network cover the whole 170 km^2 outdoor oil-field (the west–east distance is around 30 km), 24 communication nodes, including 4 mesh backbone nodes and associated 20 remote transmission units (RTUs), are deployed (shown in Fig. 4.21). The maximum throughput measured by `iperf3` in the outdoor mesh and indoor Wi-Fi network is 14.3 MB/s and 26.8 MB/s, respectively.

References

1. R. Mur-Artal, J.D. Tardós, ORB-SLAM2: An open-source slam system for monocular, stereo, and RGB-D cameras. IEEE Trans. Rob. **33**(5), 1255–1262 (2017)
2. T. Pire, T. Fischer, G. Castro, P. De Cristóforis, J. Civera, J.J. Berlles, S-PTAM: stereo parallel tracking and mapping. Rob. Auton. Syst. **93**, 27–42 (2017)
3. H. Qiu, F. Ahmad, F. Bai, M. Gruteser, R. Govindan, AVR: Augmented vehicular reality, in *Proceedings of the ACM MobiSys* (2018)
4. P. Schmuck, M. Chli, CCM-SLAM: robust and efficient centralized collaborative monocular simultaneous localization and mapping for robotic teams. J. Field Rob. **36**(4), 763–781 (2019)

5. P. Schmuck, M. Chli, Multi-UAV collaborative monocular slam, in *Proceedings of the IEEE ICRA* (2017)
6. D. Zou, P. Tan, W. Yu, Collaborative visual slam for multiple agents: a brief survey. Virtual Real. Intell. Hardw. **1**(5), 461–482 (2019)
7. Y. Ma, N. Selby, F. Adib, Drone relays for battery-free networks, in *Proceedings of the ACM Sigcomm* (2017)
8. A. W. with Robots. https://www.wired.com/story/amazon-warehouse-robots/.()
9. D. drones for industrial inspection. https://www.dji.com/products/industrial.()
10. A.J.B. Ali, Z.S. Hashemifar, K. Dantu, Edge-SLAM: Edge-assisted visual simultaneous localization and mapping, in *Proceedings of the ACM Mobisys* (2020)
11. J. Xu, H. Cao, D. Li et al., Edge assisted mobile semantic visual slam, in *Proceedings of the IEEE INFOCOM* (2020)
12. L. Liu, H. Li, M. Gruteser, Edge assisted real-time object detection for mobile augmented reality, in *Proceedings of the ACM MobiCom* (2019)
13. T.Y.-H. Chen, L. Ravindranath, S. Deng, P. Bahl, H. Balakrishnan, Glimpse: Continuous, real-time object recognition on mobile devices, in *Proceedings of the ACM Sensys* (2015)
14. F. Ahmad, H. Qiu, R. Eells, F. Bai, R. Govindan, CarMap: Fast 3D feature map updates for automobiles, in *Proceedings of the USENIX NSDI* (2020)
15. C. Cadena, L. Carlone, H. Carrillo et al., Past, present, and future of simultaneous localization and mapping: toward the robust-perception age. IEEE Trans. Rob. **32**, 1309–1332 (2016)
16. R. Mur-Artal, J.M.M. Montiel, J.D. Tardos, ORB-SLAM: a versatile and accurate monocular slam system. IEEE Trans. Rob. **31**(5), 1147–1163 (2015)
17. T. Dataset. https://vision.in.tum.de/data/datasets/rgbd-dataset/tools.()
18. K. Dataset. http://www.cvlibs.net/datasets/kitti/eval_odometry.php.()
19. E. Dataset. https://projects.asl.ethz.ch/datasets/.()
20. P. Mühlfellner, M. Bürki, M. Bosse, W. Derendarz, R. Philippsen, P. Furgale, Summary maps for lifelong visual localization. J. Field Rob. **33**(5), 561–590 (2016)
21. N.J.A. Xavier. https://developer.nvidia.com/embedded/jetson-agx-xavier-developer-kit.()
22. R. Melodic. https://wiki.ros.org/melodic.()
23. B. Triggs, P.F. McLauchlan, R.I. Hartley, A.W. Fitzgibbon, Bundle adjustment—a modern synthesis, in *Proceedings of the Springer International Workshop on Vision Algorithms* (1999)
24. D. Van Opdenbosch, E. Steinbach, Collaborative visual slam using compressed feature exchange. IEEE Rob. Autom. Lett. **4**(1), 57–64 (2018)
25. V. Opdenbosch et al., Data compression for collaborative visual slam, Ph.D. Dissertation, Technische Universität München, 2019
26. M. Dymczyk, S. Lynen, M. Bosse, R. Siegwart, Keep it brief: Scalable creation of compressed localization maps, in *2015 IEEE/RSJ International Conference on Intelligent Robots and Systems (IROS)* (2015), pp. 2536–2542. https://doi.org/10.1109/IROS.2015.7353722
27. M. Dymczyk, T. Schneider, I. Gilitschenski, R. Siegwart, E. Stumm, Erasing bad memories: Agent-side summarization for long-term mapping, in *Proceedings of the IEEE IROS* (2016)
28. S. Rusinkiewicz, M. Levoy, Efficient variants of the ICP algorithm, in *Proceedings of the IEEE 3-D Digital Imaging and Modeling* (2001)
29. B. Bescos, J.M. Fácil, J. Civera, J. Neira, DynaSLAM: tracking, mapping, and inpainting in dynamic scenes. IEEE Rob. Autom. Lett. **3**(4), 4076–4083 (2018)
30. C. Campos, R. Elvira, J.J.G. Rodríguez, J.M. Montiel, J.D. Tardós, Orb-slam3: An accurate open-source library for visual, visual–inertial, and multimap slam. IEEE Trans. Rob. **37**(6), 1874–1890 (2021)
31. F. Li, S. Yang, X. Yi, X. Yang, CORB-SLAM: A collaborative visual slam system for multiple robots, in *International Conference on Collaborative Computing: Networking, Applications and Worksharing* (Springer, Berlin, 2017)
32. Opti-Track. https://optitrack.com/.()
33. M. Karrer, P. Schmuck, M. Chli, CVI-SLAM—collaborative visual-inertial slam. IEEE Rob. Autom. Lett. **3**(4), 2762–2769 (2018)

34. S. Zhang, W. Wang, T. Jiang, Wi-Fi-inertial indoor pose estimation for microaerial vehicles. Trans. Ind. Electron. **68**(5), 4331–4340 (2020)
35. G. Klein, D. Murray, Parallel tracking and mapping for small AR workspaces, in *Proceedings of the IEEE ISMAR* (2007)
36. J. Engel, T. Schöps, D. Cremers, LSD-SLAM: Large-scale direct monocular slam, in *Proceedings of the European Conference on Computer Vision (ECCV)* (Springer, Berlin, 2014), pp. 834–849
37. R.A. Newcombe, S.J. Lovegrove, A.J. Davison, DTAM: Dense tracking and mapping in real-time, in *Proceedings of the IEEE ICCV* (2011)
38. N. Engelhard, F. Endres, J. Hess, J. Sturm, W. Burgard, Real-time 3D visual slam with a hand-held RGB-D camera, in *Proceedings of the RGB-D Workshop on 3D Perception in Robotics at the European Robotics Forum, Vasteras, Sweden*, vol. 180 (2011), pp. 1–15
39. M. Labbé, F. Michaud, RTAB-Map as an open-source Lidar and visual simultaneous localization and mapping library for large-scale and long-term online operation. J. Field Rob. **36**, 416–446 (2019)
40. R. Mur-Artal, J.D. Tardós, Visual-inertial monocular slam with map reuse. IEEE Rob. Autom. Lett. **2**(2), 796–803 (2017)
41. T. Qin, P. Li, S. Shen, VINS-Mono: a robust and versatile monocular visual-inertial state estimator. IEEE Trans. Rob. **34**(4), 1004–1020 (2018)
42. T. Wei, A. Zhou, X. Zhang, Facilitating robust 60 GHz network deployment by sensing ambient reflectors, in *Proceedings of the USENIX NSDI* (2017)
43. C.X. Lu, S. Rosa, P. Zhao et al., See through smoke: Robust indoor mapping with low-cost mmWave radar, in *Proceedings of the ACM MobiSys* (2020)
44. L. Riazuelo, J. Civera, J.M. Montiel, C2tam: a cloud framework for co-operative tracking and mapping. Rob. Auton. Syst. **62**(4), 401–413 (2014)
45. C. Forster, S. Lynen, L. Kneip, D. Scaramuzza, Collaborative monocular slam with multiple micro aerial vehicles, in *Proceedings of the IEEE IROS* (2013)
46. ArduPilot. https://ardupilot.org/ardupilot/.()
47. N. G. Acceleration. https://numba.pydata.org/.()
48. C. Toolkit. https://developer.nvidia.com/cuda-toolkit.()

Part III
Edge-Assisted Visual SLAM: Innovations and Applications

Part III delves into the practical applications of edge-assisted visual SLAM, exploring how these systems revolutionize autonomous navigation, enhance mapping and system accuracy with crowdsourcing, advance environment segmentation, and evolve multi-user augmented reality experiences.

Chapter 5 discusses pioneering strategies for visual navigation in dynamic environments, their real-time implementation, and evaluation in various scenarios.

Chapter 6 focuses on fusing maps for global accuracy and the reliable design of mobile-edge structured environments, providing insights into path planning and navigation.

Chapter 7 presents novel methods for mobile mask transfer and edge inference acceleration, aiming for fine-grained accuracy in environment segmentation.

Lastly, Chap. 8 tackles the complexities of multi-user AR, showcasing innovations in shared on-device tracking, system synchronization, and the application of body affinity filters.

Chapter 5
Indoor Autonomous Navigation with EdgeSLAM

Abstract Contemporary approaches to indoor autonomous navigation heavily depend on pre-established extensive location systems equipped with detailed maps and require a significant infrastructure setup. This chapter introduces Pair-Navi, an innovative system that bypasses these necessities by leveraging the path information shared by an initial user (i.e., the leader) to guide subsequent navigators (i.e., followers) using Peer-to-Peer (P2P) technology. Utilizing the capabilities of visual SLAM accessible on today's smartphones, our system addresses the challenges posed by environmental dynamics and the computational demands that usually limit real-time execution. We enhance system resilience against variable environmental conditions by filtering out transient elements, focusing solely on stable, unchanging features. For seamless real-time operation on handheld devices, Pair-Navi strategically segregates and rearranges the intertwined components of SLAM for both leaders and followers. Demonstrated on widely available smartphones across varied architectural environments and tested against two renowned benchmarks (TUM and KITTI), Pair-Navi showcases a near-instant navigation success rate of 98.6%, with effectiveness slightly reduced to 83.4% after 2 weeks, significantly surpassing existing alternatives by over 50%. As a genuinely infrastructure-independent solution, Pair-Navi illuminates the path forward for efficient and practical indoor navigation for mobile consumers.

5.1 Introduction

The concept of Peer-to-Peer (P2P) navigation has emerged as a novel solution to bypass the need for installing indoor localization infrastructures [1]. In this model, an initial explorer, termed as *leader*, compiles navigational data (e.g., directional changes and key environmental features) and disseminates this information via the Internet to subsequent *followers* embarking toward the same goal. An illustrative use case is a merchant-generated navigation service guiding customers within a large shopping complex directly to a specific store, thus obviating reliance on third-party location services.

J. Xu et al., *Edge Assisted Mobile Visual SLAM*,
https://doi.org/10.1007/978-981-97-3573-0_5

Innovative efforts have explored this leader–follower scheme for P2P wayfinding [1–3], utilizing environmental Wi-Fi signals alongside smartphone's inertial sensors and camera snapshots as navigational cues to align the pathways of leaders and followers. Despite their ingenuity, the dependence on Wi-Fi installation and inaccuracies due to the fluctuating Wi-Fi signal strength and sensor variability render these solutions suboptimal for real-world application.

Two recent advancements, however, promise to address these challenges and pave the way for a viable indoor navigation system. Firstly, the field of simultaneous localization and mapping (SLAM) has seen remarkable progress, with visual SLAM advancements enabling its application on standard smartphones, which typically feature a single rear camera [4–6]. Secondly, the enhancement of visual processing capabilities on mobile platforms now supports sophisticated vision-based tasks, further bolstering mobile-based navigation potential.

In this chapter, we investigate visual SLAM with the power of mobile vision and present Pair-Navi, a P2P indoor navigation system that requires no pre-existing or dedicatedly installed infrastructure, pre-deployed localization service, or indoor digital maps. Visual SLAM utilizes one or more cameras to explore an unknown environment by continuously locating the camera itself in the environment and meanwhile constructing a map of the environment [6]. Our approach is built upon monocular visual SLAM with a single camera commonly equipped on commodity smartphones. A leader of Pair-Navi simply walks through a path recording a video clip along the route and shares the trajectory video via a cloud server for potential upcoming followers. Pair-Navi consumes the video for SLAM and constructs the trajectory that the leader has traveled. When a follower appears, he/she will be provided with a leader's trajectory as reference. On the follower side, Pair-Navi also captures real-time video frames and precisely locates the follower's relative location to the reference trajectory, accordingly navigating the follower by timely promoting walking hints. As shown in Fig. 5.1, both the current scene and the reference trace will be displayed to the follower. In addition, Pair-Navi handles navigation deviation, which is necessary but neglected by existing P2P navigation approaches. Without the need to instrument the building of interests, Pair-Navi works in any scenes as long as a camera-equipped smartphone is available.

Adapting visual SLAM to a dependable navigation framework faces significant hurdles: (1) *Dynamic Environments*. Conventional SLAM methodologies typically operate under the assumption of static environments, anticipating no alterations during or after map development [5–7]. Yet, real-world settings are subject to continuous change, with elements like pedestrian movement, shifts in furniture placement, and variability in lighting conditions, which compromise the fidelity of mapped trajectories. As depicted in Fig. 5.2, this dynamic nature can lead to inaccuracies in frame alignment, substantially affecting SLAM performance. Addressing these dynamic challenges in settings abundant with changes, such as bustling malls or airports, proves difficult with existing visual SLAM techniques [6, 8]. (2) *Immediate Response*. The demand for high computational power for key operations such as odometry and map optimization often exceeds what is feasible on standard mobile devices, impeding the ability to deliver real-time navigation

Fig. 5.1 Follower's navigation interface of Pair-Navi

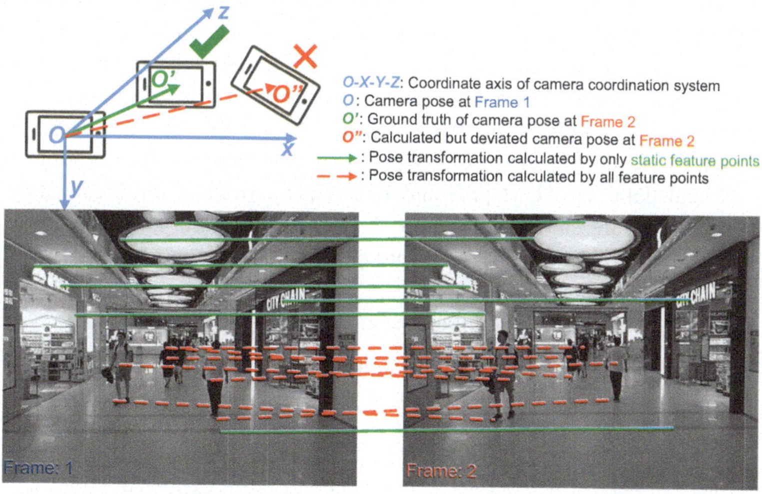

Fig. 5.2 Illustration of feature point matching of two consecutive frames. Feature points from rigid contexts are matched correctly (solid green lines), while most feature points from non-rigid contexts are mismatched (dashed red lines), resulting in errors in camera pose calculation

solutions. An ideal navigation tool must not only accurately determine the user's position and plot the route instantly but also present navigational cues in a user-friendly manner on the fly.

To counteract the issue of dynamic environments, our approach involves isolating and removing the variable elements, such as moving individuals and transient objects from the video data, focusing solely on stable aspects for SLAM processes. This strategy is grounded in two main observations: firstly, the abundance of stationary features in standard indoor scenes allows for selective filtering of dynamic elements without compromising SLAM effectiveness. Secondly, advancements in computer vision, particularly deep learning innovations, enable precise detection and segmentation of transient entities within videos [9–11]. Leveraging Mask R-CNN [11], we distinguish and exclude these dynamic objects from the SLAM process, thus mitigating the influence of environmental changes and preserving the integrity of navigation paths.

For real-time navigation enablement on smartphones, our system, Pair-Navi, simplifies and reorganizes the interconnected SLAM components, focusing only on those essential for follower guidance. Instead of utilizing a full SLAM setup for followers, Pair-Navi opts for relocalization to align the follower's path with the leader's recorded trajectory. Additionally, a synchronization approach is employed to efficiently manage the visual navigation and dynamic content removal tasks. This method ensures that Pair-Navi facilitates swift and accurate navigational assistance on mobile platforms, unlike other recent endeavors that incorporate semantic elements into SLAM but struggle with real-time performance [12–14].

We implement our system on the Robot Operating System (ROS) platform [15] on the server and on ROS-Android [16] on the phone side. Comprehensive experiments are carried out in three buildings with various conditions over 2 weeks. The results demonstrate that Pair-Navi achieves a remarkable navigation success rate of 98.6%. Even after 2 weeks since the construction of the leaders' trajectory, the rate maintains 83.4%, outperforming the state-of-the-art *Travi-Navi* [3] by 50.9% and *FollowMe* [2] by 80.4%. We further evaluate the efficiency of NRCC in two official datasets: TUM [17] and KITTI [18]. Both of the localization and mapping accuracy are increased by more than 25%. Being truly infrastructure-free, Pair-Navi takes an important step toward practical indoor navigation for mobile users.

In summary, the core contributions are as follows:

- We present the first user-friendly vision-based P2P navigation system, which neither requires to instrument a building nor relies on pre-deployed localization service with indoor maps.
- We employ non-rigid context culling by using Mask-RCNN to overcome indoor environmental dynamics for visual SLAM, which significantly improves the robustness and precision of navigation. Extensive experiments on two standard datasets demonstrate the leverage of NRCC improves the localization accuracy and mapping precision.
- We implement a complete real-time system on commodity smartphones and extensively evaluate the performance. The results show Pair-Navi achieves delightful results and outperforms all existing solutions.

The rest of chapter is organized as follows. We present an overview in Sect. 5.2 and introduce visual navigation in non-rigid environment in Sect. 5.3. Real-time design and implementation is provided in Sect. 5.4, followed by experiments in Sect. 5.5. We review related works in Sect. 5.6 and conclude in Sect. 5.7.

5.2 EdgeSLAM-Based System Architecture for Autonomous Navigation

5.2.1 Peer-to-Peer Navigation

In contrast to traditional navigation systems dependent on established localization infrastructures, Pair-Navi operates through a straightforward P2P navigation framework. This method eliminates the requirement for detailed indoor maps, which can be challenging to acquire and manage. The P2P system comprises two primary participants: a leader, who shares their navigational experience, and a follower, who uses this shared guidance. Essentially, any individual traversing a route can act as a leader for that path by sharing specific navigational data (e.g., sequences of Wi-Fi signals, geomagnetic patterns, and readings from IMU sensors [1–3], or in our implementation, video recordings) along with automatically generated directional cues (e.g., orientation adjustments, ascents, descents, etc.). This data is subsequently provided to followers to aid in aligning their movements with the leader's recorded path, allowing them to navigate accordingly. Importantly, an individual's role as either leader or follower is adaptable based on the situation at hand.

P2P navigation shines as a quick and effortless deployment option for regular users looking to offer localized navigational assistance. For instance, a retailer might create a bespoke navigation guide to lead customers directly to their store, or a conference organizer could easily direct participants to event venues. In many potential applications, P2P navigation presents a viable, low-effort alternative to conventional, centrally managed navigation and localization solutions.

5.2.2 System Overview

Pair-Navi harnesses mobile vision to facilitate a novel leader–follower navigation method. The system's blueprint, depicted in Fig. 5.3, requires participants to simply record their journey using smartphone cameras. Each frame captured is transmitted to a cloud server for processing. Unlike traditional SLAM applications that utilize all components for both roles, Pair-Navi selectively engages specific SLAM modules based on the user's role. For leaders, the system processes video through three pivotal SLAM modules: *Initialization*, *Visual Odometry*, and *Trajectory Construction*, crafting both a navigational path and a spatial map. Upon completion, this data, alongside marked locations, is archived on the server.

Upon a follower's entry, choosing a destination triggers the retrieval of a prior leader's route to the same endpoint. As the follower progresses, Pair-Navi accurately gauges their position relative to the predetermined path via *relocalization* from current footage. Successful relocalization ensures the follower receives timely navigational prompts derived from the leader's path. Conversely, the detection of potential deviation activates a *deviation detection* mechanism, employing visual

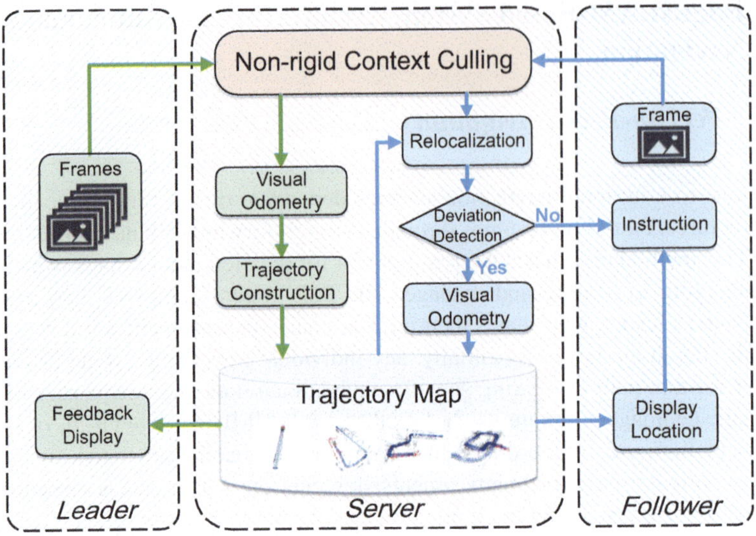

Fig. 5.3 Pair-Navi architecture

odometry to independently ascertain the follower's orientation. This data, in tandem with the original route, aids the follower in recalibrating their direction.

An innovative feature within Pair-Navi is the *Non-Rigid Context Culling* (NRCC), designed to isolate and remove transient elements from video content to navigate dynamic settings effectively. By applying NRCC to both leaders' and followers' recordings, Pair-Navi guarantees the generation of accurate routes and maintains reliable localization for followers.

5.3 Navigation in Non-rigid Environment

In this section, we describe how Pair-Navi utilizes visual SLAM for navigation and how it addresses non-rigidity in the environment.

5.3.1 Visual SLAM for Navigation

In Pair-Navi, we use monocular visual SLAM that works with a single camera since most smartphones have only one camera on the back side. We decouple the tightly coupled modules of a monocular visual SLAM system and reorganize the required modules into leader and follower applications to simplify for real-time meanwhile ensuring accuracy. In particular, we introduce three key modules, initialization, visual odometry, and relocalization, as follows:

Initialization When the system initially launches, our system takes an initialization step using epipolar geometry [6, 19] to locate the camera in an initial map with 3D map-points as landmarks of the environment. To begin with, the camera captures two very early video frames, from which we extract and match certain feature points that describe the video frames (see Fig. 5.5a). In Pair-Navi, we leverage ORB feature point [20], which has comparative matching accuracy with the state-of-the-art SIFT [21] and SURF [22], yet far more efficient in computation.

Then the relative camera motion between the first two frames, denoted as ${}^{c_1}_{c_2}T$, where T represents a transformation matrix and c_i denotes certain coordinate systems,[1] is computed by solving the so-called epipolar constraint equation [19]. When the camera motion between the early two frames is acquired, the camera pose of the first frame is set to be the "world coordinate system" w, and the camera pose of the second frame becomes ${}^{w}_{c_2}T$. Based on camera motion between them, the depths of the feature points are also calculated by triangulation [19]. After triangulation, we obtain the 3D coordinates (2D pixel coordinate plus depth) of every feature point in the camera coordinate system. Since the first frame is set to be the world coordinate, we actually get the 3D coordinates of every feature point in the world coordinate. Therefore, an initial sparse 3D map of the environment, i.e., a sparse set of 3D map-points, is accordingly built up.

Visual Odometry After initialization, the visual odometry (VO) module takes charge of the system, to continuously track the camera pose from consecutive video frames. Specifically, when a new video frame arrives, its 2D ORB feature points are extracted and associated with already-created 3D map-points by feature matching.

Figure 5.4 shows an example of the extracted feature points associated with 3D map-points. From the association of 2D feature points and 3D map-points, we can acquire the camera pose of this frame by solving a so-called Perspective-n-Point (PnP) problem [24], which determines the position and orientation of a camera from a set of correspondences between 3D points and 2D pixel points. As shown in Fig. 5.4, given the camera poses of two frames, new 3D map-points can be generated by calculating the 3D coordinates via triangulation among the two frames [19]. Repeatedly, as more new frames come, a trajectory of the camera poses and a map of the 3D landmarks and corresponding keyframes[2] are built incrementally. In Pair-Navi, VO is the core module for a leader to construct a trajectory map.

Relocalization In order to reuse a previously built trajectory map, relocalizaiton module comes in handy, which is the central component in the follower program. It compares a video frame with the keyframes in the map and finds out the most

[1] This convention is brought up by Craig [23], where the left superscript of a variable means reference coordinate system and the left subscript means the objective coordinate system. By multiplication, we can cancel the left subscript of a variable together with the left superscript of the next variable. For example, ${}^{A}t = {}^{A}_{B}T \cdot {}^{B}t$ means a vector t described in B coordinate system multiplied by transformation ${}^{A}_{B}T$ equals the description of t in A coordinate system.

[2] Keyframes are a subset of all frames to eliminate redundancy.

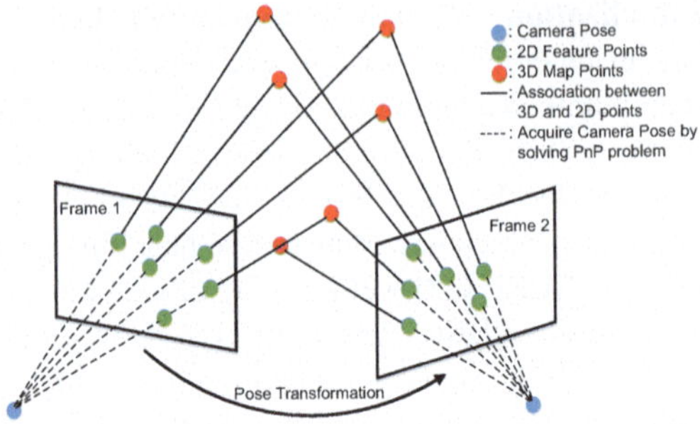

Fig. 5.4 Illustration of visual odometry

similar keyframe based on feature point matching. This step is also called visual place recognition, of which the state-of-the-art is Bags of Visual Words [25]. After the most similar keyframe is found, the feature points in the current follower frame are associated with the feature points in the selected keyframe. On this basis, a PnP problem [24] is solved in the same manner as VO module to get the camera pose, thus relocalizing the camera in the map. In Pair-Navi, the follower program mainly employs relocalization to achieve efficient relative localization (Sect. 5.4).

Note that classical monocular visual SLAM still involves complicated steps like loop closing detection, global optimization, etc. [6]. In our system, however, we merely apply the necessary modules for leaders and followers, respectively, to avoid intensive computation, as detailed in Sect. 5.4.

5.3.2 Non-rigid Context Culling (NRCC)

5.3.2.1 Limitations of Visual SLAM in Non-rigid Environments

While visual SLAM technology underpins a promising solution to infrastructure-free navigation, it is vulnerable to non-rigid indoor environments with significant dynamic changes over time. Specifically, the limitations are twofold:

Low-Precision Trajectory As is described in the above section, to calculate the camera poses, we need to match feature points first. Therefore, correct feature point matches influence the accuracy of the constructed trajectory. In the presence of erroneous matches, the generated trajectory will deviate from the ground truth. Figure 5.2 illustrates an example of matching two consecutive video frames from a typical shopping mall. As seen, feature points extracted from those dynamic contexts (e.g., pedestrians) will lead to considerable erroneous matching, as indicated by red lines in Fig. 5.2. As a consequence, if we calculate camera poses from the whole set of feature point matches without screening, the constructed trajectory will deviate from the truth, depraving further navigation. To obtain precise trajectory, we need to intelligently recognize the non-rigid contexts and sift out their corresponding feature points.

Vulnerable Relocalization Apart from degrading trajectory precision, non-rigid contexts further harm relocalization robustness. In practice, the environment observed by a leader and later a follower may change significantly, leading to feature point mismatches and thus large relocalization errors or even relocalization failures. In one situation, if there are only a small number of matching outliers, the feature points may be matched to wrong 3D map-points, resulting in errors in camera pose computation. In another situation, if a large portion of feature points fail to match, a wrong keyframe will be chosen and relocalization fails.

Therefore, to ensure accurate trajectory construction for leader and successful relocalization for follower navigation, we propose the non-rigid context culling (NRCC) module that takes out features points from non-rigid contexts and only exploits the remaining reliable feature points, mainly from rigid and static areas, for frame matching. Our key observation is that there are abundant feature points for frame matching, allowing room to cull part of them without degrading visual SLAM performance.

5.3.2.2 Non-rigid Context Culling via Mask R-CNN

In Pair-Navi, we adapt Mask R-CNN [11] for NRCC. Mask R-CNN is a recent framework for instance segmentation. It aims to separate different instances in an image via a segmentation mask for each instance. We use the Mask R-CNN pretrained on COCO dataset [26] and select the object categories that are suitable for indoor scenario.

Since we aim at distinguishing rigid and non-rigid context, we divide all the object categories into two sets: rigid context objects and non-rigid context objects as shown in Table 5.1. If an object belongs to the rigid context set, it means the location, pose, and shape of the object will not change, and whenever the leader or follower comes to the same place, they will observe the object in the same situation. On the contrary, an object is dynamic if it belongs to the non-rigid context set. What is worth mentioning is the classification of the two sets is flexible. Some objects (e.g. vase, potted plant) can belong to either the rigid context object set or the other, depending on the time interval between the leader's trajectory being constructed and the follower's navigation. For example, for vases, if the time interval is shorter than 7 days, they will be regarded as rigid context objects. Yet if the interval is longer than 7 days, they will be treated as non-rigid context.

When the server receives a video frame from the camera, we extract its feature points and use Mask R-CNN framework to detect the non-rigid contexts, and then we filter out the feature points that lie in the masks of dynamic instances. As shown in Fig. 5.5, the feature points filtered out, which are marked by red color, are mainly from people, and the map is generated only using the feature points belonging to static environment, which are marked by green. Moreover, we also use the method proposed in YOLO [27] to detect mirrors and smooth surfaces in video frames (as shown in Fig. 5.5b). We believe the illumination change in places like academic buildings may be drastic, rendering the feature points lying in mirrors

Table 5.1 Object classification for indoor scenario

Non-rigid context objects	Rigid context objects
Person, hair drier, toothbrush, cat, keyboard, phone, apple, cup, backpack, umbrella, handbag, tie, suitcase, dog, book, clock, skis, snowboard, sports ball, kite, bat, skateboard, surfboard, mouse, remote, glass, fork, knife, bowl, banana, tennis racket, scissors, teddy bear, sandwich, broccoli, carrot, hot dog, pizza, donut, cake, chair	Couch, potted plant*, dining table TV, microwave, oven, toaster refrigerator, vase*, bed, toilet, sink

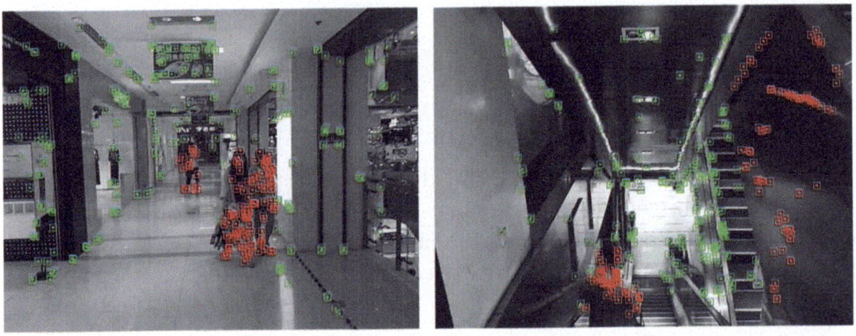

(a) (b)

Fig. 5.5 Examples of non-rigid contexts: feature points from non-rigid contexts, pedestrians in (**a**) and mirror reflections in (**b**), are recognized and marked as red, while the remaining feature points from rigid and static objects are green

volatile. Therefore they are also culled to increase system robustness. After this preprocessing of the video frame, although the user may be facing a non-rigid environment, the feature points lying in the masks of dynamic instances will not be involved in trajectory construction (for leaders) or relocalization (for followers). As shown in Fig. 5.6, removing non-rigid contexts helps improve precision and avoid potential relocalization failures.

5.4 Real-Time Design for Autonomous Navigation

In this section, we present the design of Pair-Navi to enable real-time navigation.

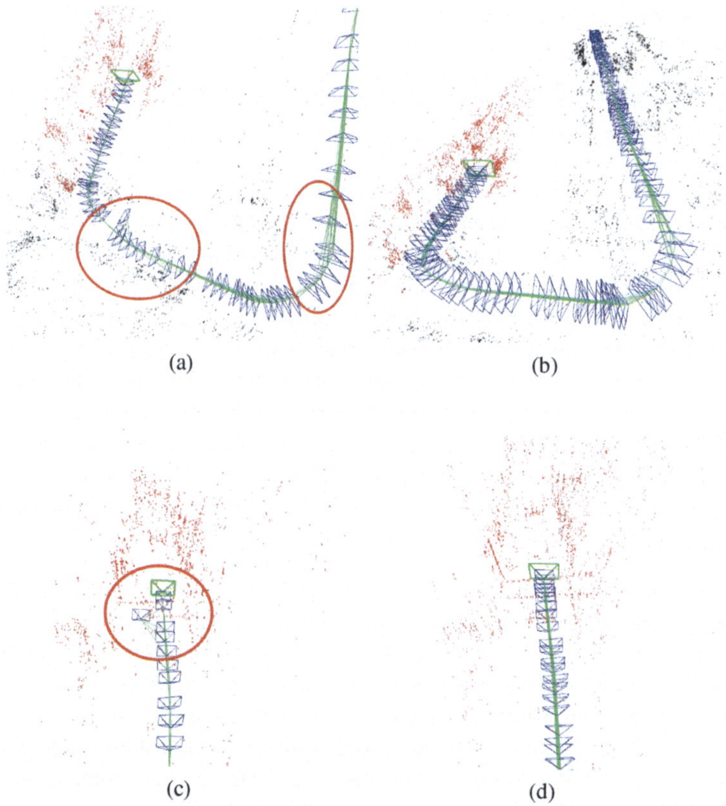

Fig. 5.6 The effect of NRCC on trajectory construction: a complex trajectory computed (**a**) without NRCC, in which the camera poses of some frames in the red circle are deviated from the original path, and (**b**) with NRCC, in which the camera poses are correct and smooth. Similarly, (**c**) and (**d**) show a simple trajectory without and with NRCC, respectively

5.4.1 Leader Trajectory Map Construction

To construct a map of navigation trajectories, a leader simply holds the smartphone in front of his/her body to shoot video frames while walking along his path. Upon finishing, the leader uploads to the server the video labeled with the origin and the destination. The server then runs visual SLAM, including initialization and visual odometry, with Mask R-CNN for each frame to cull out the non-rigid contexts, and then the map is saved for followers.

5.4.2 Follower Real-Time Navigation

To begin with, a follower chooses the destination given by the server, and then the smartphone runs as an ROS node, capturing video frames and sending them to the server. The server, as another ROS node, loads the chosen trajectory and map and starts running: It relocalizes every follower video frame, visualizes the camera pose in the leader's trajectory map, and calculates the navigation instruction. On receiving all the returned messages, the follower can see the navigation instruction, together with a visualization of the current camera pose and the leader's trajectory map.

If the relocalization observes inadequate quantity of 3D map-points, the auxiliary VO is launched from this frame and will take place of the relocalization to show camera pose if it truly fails. At this time, the follower can still see the camera pose in the leader's trajectory and spare little effort to return to the track. Once the relocalization is successful again, the auxiliary VO is shut down, and the follower goes back to the normal case, in which only relocalization is executed.

Our system has the following three designs that ensure it can run in real-time (by "real-time," we mean at least 10 FPS, the typical value of persistence of vision):

Relocalization We decouple a typical visual SLAM system, adaptively combining the relocalization module and VO module into our follower navigation program. In this design, we avoid the heavy overhead to acquire the follower's current position with a full visual SLAM system since the beginning of the follower's navigation. Instead, we get the utmost of the leaders' efforts, only running relocalization for the follower, which is more efficient than a full visual SLAM system yet produces as accurate results as it. As will be demonstrated in Sect. 5.5, our design considerably reduces running latency compared to a conventional SLAM system.

Mask Synchronization Strategy Non-rigid context culling is for both leader and follower's video frames, but in different ways. For leaders, we generate one mask for each frame, because the leader's map construction can be done offline. Yet for followers, the calculation of Mask R-CNN is relatively slow, at an average frame rate of about 5 FPS. So we take a trade-off strategy to synchronize Mask R-CNN with visual SLAM. We aggregate every two frames (corresponding to 0.2 s due to 10 FPS) and generate one mask for them both.

The rationale behind is that a user will move for only about 20 cm during the 0.2 s time-window, assuming a typical walking speed of 1 m/s. Therefore the potential scene changes in two consecutive frames will not be too significant, which may slightly affect trajectory construction accuracy for leaders but do not necessarily influence relocalization for followers. In summary, to obtain a highly precise and reliable trajectory for followers to use, we need to cull non-rigid contexts for each frame; yet to save the computational resources for real-time follower navigation, we can confidently reuse the masks for several consecutive frames.

The Fringe Benefit of NRCC Additionally, NRCC brings some fringe benefit to the real-time performance. Since the number of feature points is reduced, further computation, including feature point matching, calculating bags of visual words,

PnP, etc., is simplified by the proportion of the rigid feature points in all the feature points. The quantitative benefit of this simplification is shown in Fig. 5.15.

5.4.3 Follower Deviation Handling

Existing P2P navigation systems such as *Travi-Navi* and *FollowMe* have no function of deviation handling, and thus if follower deviation happens, the tracking will lose and the navigation will fail, with no recovery approach. In contrast, we add an auxiliary VO module to track the deviated trajectory, which acts as a fail-safe measure to guarantee the success of navigation.

In our scenario, deviation happens when the follower walks off the instructed course or even when camera pose slightly deviates. In these cases, the follower's relocalization module observes inadequate 3D map-points in a frame, which is when the system launches the auxiliary VO. The auxiliary VO, extracting new feature points and generating new 3D map-points, runs in parallel with the relocalization to retain continuous tracking and show the camera pose in the trajectory map, as shown in Fig. 5.7. With the camera pose shown in the trajectory map, the follower can easily go back on the right track. Once the relocalization observes a healthy quantity of 3D map-points, the auxiliary VO will shut down and handover the navigation to the relocalization module.

Note that to further keep our system running in real-time with this auxiliary VO, we strictly select a feature point quantity threshold for triggering it, so that redundant calculation is eliminated. In this way, we make the system tolerate navigation deviations in harsh environments while still maintaining real-time performance.

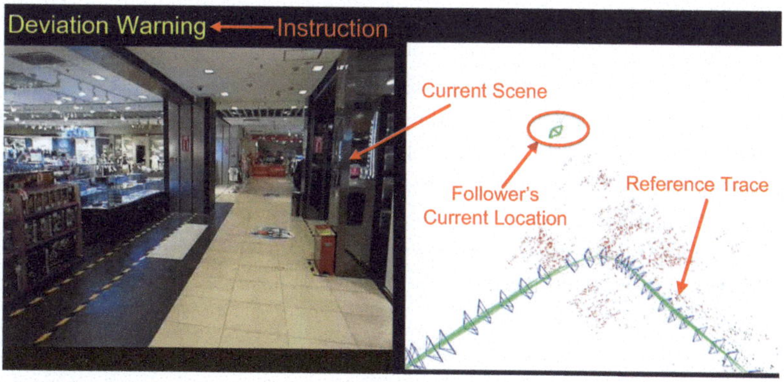

Fig. 5.7 Illustration of auxiliary VO when deviation happens

5.4.4 Navigation Instruction Calculation

The calculation of navigation instruction takes the relocalized camera pose of the current follower frame with reference to the world coordinate system, ${}^{w}_{c}T$, as an input (the left superscript w denotes "world," which is set to be the first frame of the leader's map, and the left subscript c denotes "current"). In this way, every follower frame is localized and assigned a pose in the leader's trajectory map. Next, the server needs to compute navigation instruction through the follower's current camera pose. Since in the leader's trajectory, there is one pose ${}^{w}_{r}T$ (r denotes "relocalization") that is the closest to the follower's current camera pose, we choose it as a representative and average the translation[3] of the next ten poses in the leader's trajectory as ${}^{w}t_{next}$. The reason why we select following ten poses is that the average time interval of ten keyframes is approximately 1.5 seconds by our statistic, and the average step period of human is around 0.4~0.6 s; therefore, this ten-keyframe interval suggests approximately in which direction the follower should head to in the next two or three steps.

So far, we have the pose of the current follower frame ${}^{w}_{c}T$ and the average translation of the next ten poses ${}^{w}t_{next}$. By converting t_{next} to homogeneous coordinate (adding a fourth dimension with value 1, such that t_{next} becomes $[x\ y\ z\ 1]^T$), we can compute

$$
{}^{c}t_{next} = {}^{c}_{w}T \cdot {}^{w}t_{next}, \tag{5.2}
$$

where ${}^{c}t_{next}$ is the homogeneous coordinate of t_{next} in the current follower frame. The coordinate axis direction is already shown in Fig. 5.2.

Finally, we can generate the navigation instruction with the angle between ${}^{c}t_{next}$ and the imaging plane of the camera

$$
\alpha = \arctan \frac{z}{x}, \tag{5.3}
$$

where we assume the follower is holding the phone in a steady level so that the y coordinate of ${}^{c}t_{next}$ has little effect on navigation.

As shown in Fig. 5.8, by our design, if $\frac{\pi}{3} < \alpha < \frac{2\pi}{3}$, the system will give out "Go Straight" instruction; when $0 < \alpha < \frac{\pi}{3}$, "Turn Right"; and when $\frac{2\pi}{3} < \alpha < \pi$, "Turn Left." Note that we assume the leader will not go backward when constructing the trajectory map, and thus theoretically z is always positive, so no "Go Backward"

---

3 Here the translation is a 3D vector denoting the position of a pose. A pose can be denoted by a transformation matrix

$$
T = \begin{bmatrix} R & t \\ \mathbf{0} & 1 \end{bmatrix}, \tag{5.1}
$$

where R is 3×3 rotation matrix, and t is the translation vector.

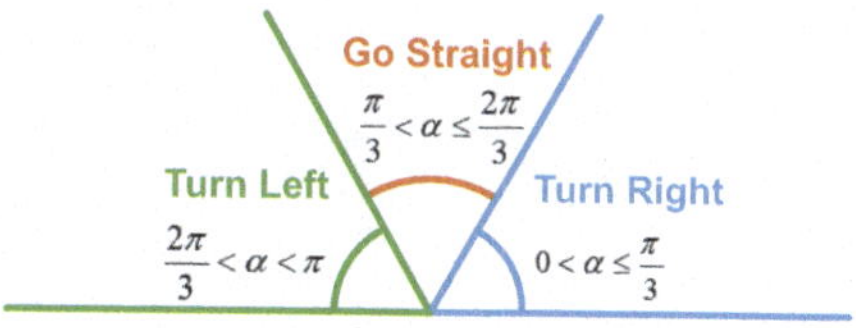

Fig. 5.8 Illustration of relationship between navigation instruction and α

instruction is necessary. If z is truly negative, which is due to flawed relocalization, we will give "Deviation Warning" instruction. If a deviation is later detected, the auxiliary VO will be triggered to continue the navigation.

5.4.5 Implementation

We implement our system on ROS Kinetic [15] platform. The user program is developed on ROS-Android [16] platform. The server's SLAM program is developed on *ORB-SLAM* [6] on ROS, and we develop the system visualization upon the visualization of *ORB-SLAM*, with some modifications via OpenCV. We resize all the frames to 640×480, which is relatively high resolution and still not too large to prolong the process time of each frame. The phone-server communication resorts to ROS topic publication and subscription, which guarantees all the information is integral because ROS data transfer is based on TCP protocol.

We applied our Mask R-CNN models with the ResNet-FPN-50 backbone, and the network parameters are pretrained on COCO image dataset. The Mask R-CNN code is implemented in Python 3.6.5 with PyTorch 0.4.0.

5.5 Evaluation for EdgeSLAM-Based Autonomous Navigation

5.5.1 Experiment Settings and Methodology

Experiment Venues We conducted extensive experiments in an office building, a gymnasium, and the first–fourth floors of a shopping mall, with area sizes of about 400 m^2, 1,000 m^2, and 6,000 m^2, respectively. The three testing environments have diverse conditions. In particular, the crowded shopping mall is the most dynamic. The office building has the most drastic illumination oscillation during a day. The gymnasium has the medium crowdedness among the three environments.

Data Collection Overall, we design 21 navigation paths, including 6 short paths ($\leq$ 100 m), 7 medium paths (100–200 m), and 8 long paths ($\geq$200 m), covering all the main pathways of the testing areas. Figure 5.9 shows four trajectories of different lengths constructed from the three areas. The reference trajectories for

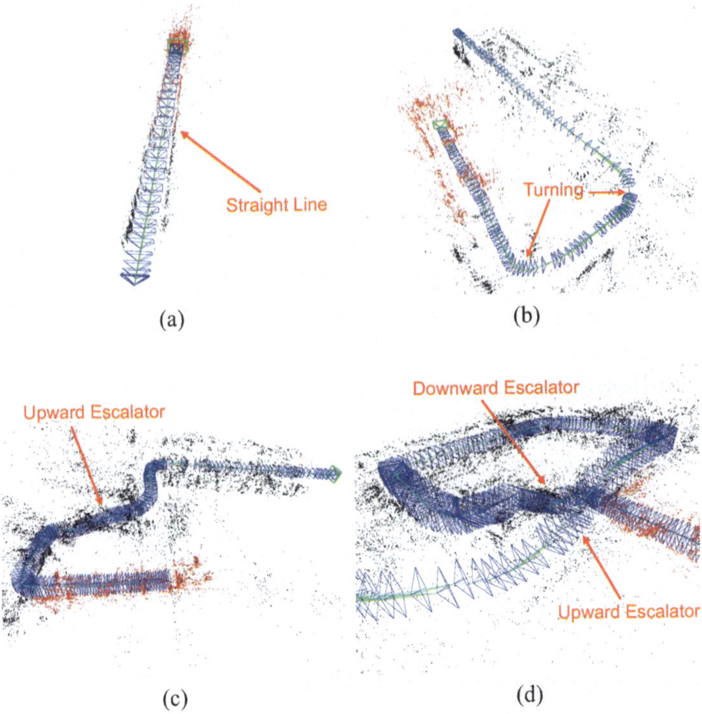

Fig. 5.9 Four typical real trajectories in our experiment: (**a**) a straight line, (**b**) a U-turn, (**c**) a complex trajectory going up an escalator, and (**d**) a complex trajectory going up one escalator and then down another

these paths are constructed by three different leaders. The total length of leaders' reference trajectories is about 3.1 km with 33,452 video frames, among which 6,781 keyframes are selected, and the followers' total walking distance is around 15 km.

Devices We tested Pair-Navi on a variety of Android mobile devices, including Huawei P10, Nexus 6p, Nexus 7, and Lenovo Phab2 pro. Since the main device discrepancies are camera intrinsics (i.e., focal length, lens center and distortion), we calibrate camera intrinsics of the smartphones and accordingly rectify video frames. The server, Lenovo IdeaPad-Y700 with i7-6700HQ CPU of 2.6 GHz main frequency and 8GB RAM, runs the Ubuntu 16.0.4 operating system and ROS Kinetic. For Mask R-CNN, the GPU we used is TITAN V with CUDA version 9.1.85 and cuDNN-7.05.

Users We recruited four volunteer followers to walk along different routes naturally as they usually do. The follower behaviors are diverse in camera holding gestures and heights. For example, one user prefers to hold the camera with two hands, while others tend to use their right hands. Thus the cameras suffer various extents of shake when the users are walking, which may cause different feature point matches. The user study is conducted on three particular days: the same day as the trajectories were constructed, 1 week later, and 2 weeks later.

Comparison To evaluate the performance of Pair-Navi, we implemented *Travi-Navi* [3] and *FollowMe* [2], two start-of-the-art P2P navigation systems for comparison.

Evaluation Metrics Similar to some existing works like *Travi-Navi* and *FollowMe*, we set checkpoints at turns, escalators, and some landmarks on each trajectory. In total, we set 274 checkpoints for the 21 navigation paths. The followers were not informed of navigation routes, the final destination, or the checkpoint locations. *Navigation success rate* is defined as the rate of successful arrival at each checkpoint in *Travi-Navi* and *FollowMe*. Thanks to the employment of deviation handling, this rate is always 100% in Pair-Navi, which means the followers arrived at the destinations successfully in all cases. So instead, we use a more strict definition of navigation success rate as $1 - p$ (where p is the rate of auxiliary VO launches) for Pair-Navi, while it keeps the original definition of navigation success rate unchanged for *Travi-Navi* and *FollowMe*.

5.5.2 *Overall Performance*

5.5.2.1 Performance Comparison

The performances of Pair-Navi and the two state-of-the-art approaches to compare are depicted in Fig. 5.10. We find that Pair-Navi achieves the best performance among all three of them, no matter how long the time interval is. The average navigation success rates by Pair-Navi in the same day of the trajectory's construction, after 1 week, and after 2 weeks are 98.6%, 93.2%, and 83.4%, respectively. Compared with the immediate performance (tested in the same day), the navigation success rates after 2 weeks decline to 14.1%, 49.3%, and 59.3% in Pair-Navi, *Travi-Navi*, and *FollowMe*. In contrast to *Travi-Navi* and *FollowMe*, Pair-Navi attains high navigation accuracy after 2-week interval and outperforms *Travi-Navi* by 50.9% and *FollowMe* by 80.4%.

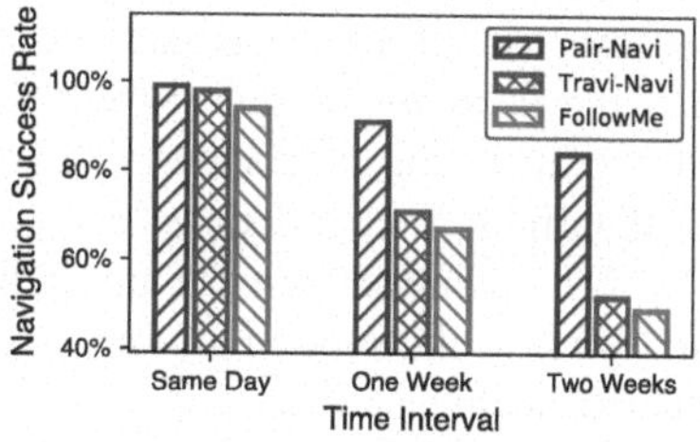

Fig. 5.10 Different methods

The reason of this performance gain is twofold:

1. Innate metric advantages of vision-based methods. On the one hand, radio frequency-based and magnetic field-based systems suffer from metric error of localization typically averaging 3~5 m [28, 29] due to intrinsic defects such as fluctuation of signal strength, change of indoor multipath environment, and the inaccuracy of pedestrian dead reckoning (PDR). On the other hand, vision-based methods tackle metric error rationally, which consists of three parts: rolling shutter cameras in most smartphones skewing the image, which is negligible, the calculation of epipolar geometry and Perspective-n-Point, which have closed form solution and is accurate, and feature point mismatch, which is largely tackled by ORB feature. As is reported in [6], typical localization error in indoor scenarios (TUM RGB-D dataset) in visual SLAM systems is within 5 cm, much smaller than that in radio frequency-based and magnetic field-based systems. With an error of 3–5 m, a person can miss turning points in indoor environments every now and then, but it is not the case with an error of 5 cm.
2. Robust design of NRCC. Apart from the innate metric advantages of vision-based systems, benefiting from the design of NRCC, the system robustness is remarkably enhanced. Radio frequency-based and magnetic field-based localization methods often struggle with ambiguity when multiple signal patterns from different locations overlap. In such scenarios, tracking mobile devices can become unreliable, as it is easy to lose track of their precise location. With the design of NRCC, our system obtains good relocalization robustness.

Furthermore, as for long-term performance, NRCC attacks environment change by ruling out the changed while maintaining the unchanged contexts in the trajectory map. In this way, even after 2 weeks, our system can handle time-varying environment and palliate the performance drop even in the long term. On the contrary, the deterioration of trajectory information in radio frequency-based and magnetic field-based systems is not dealt with properly. In Sect. 5.5.3, we will further delve into the effectiveness of NRCC.

5.5.2.2 Performance Under Different Conditions

We invite four volunteers to examine the robustness and practicability of Pair-Navi in different areas and at different times. As shown in Fig. 5.11, Pair-Navi achieves an average navigation success rate of more than 85% for each user, and the decrease of success rate for each one is less than 15% after 2 weeks. Furthermore, Fig. 5.12 shows that Pair-Navi yields a similar performance regardless of the different crowdedness levels and illumination conditions at different areas. Pair-Navi achieves consistently delightful success rates of more than 90%, 85%, and 80% in the office building, gymnasium, and large shopping mall under time interval within 2 weeks.

To further demonstrate the applicability of Pair-Navi, we note that the trajectories in the office building are mainly constructed by User-2 and User-3 and in the

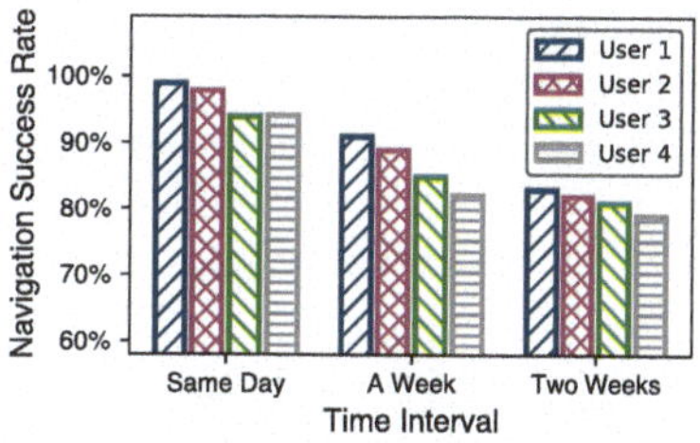

Fig. 5.11 Different users and time interval

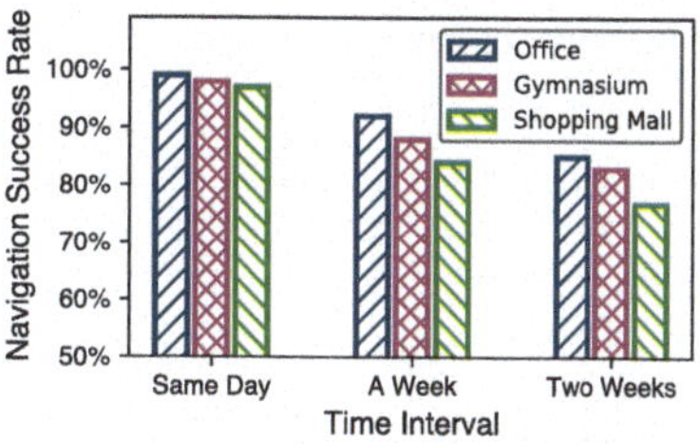

Fig. 5.12 Different areas and time interval

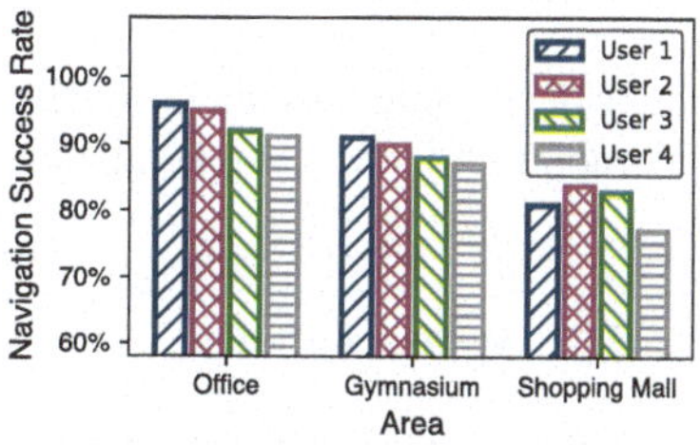

Fig. 5.13 Different users and areas

gymnasium and shopping mall by User-1 and User-2. The heights of the four users are different, and camera holding gestures are variant. Figure 5.13 reflects the robustness of Pair-Navi for different leaders and followers. Navigation success rates for all users in different areas are more the 80%, and the success rate gaps between different users are within 5%.

5.5.2.3 Impact of Illumination

We further tested our system with different illumination conditions in the office building, which undergoes the most drastic illumination oscillation among the three areas. We first asked a leader to walk four pathways in the morning, afternoon, and evening during a day. Then we asked volunteers to walk the same pathways correspondingly and calculate the navigation success rate of each test case. As shown in Fig. 5.14, whenever followers walk the pathways, the navigation success rates are more than 80% and more than 90% if followers walk in the evening. Generally, in the office building, we usually turn on all of the lights in evening, majority in morning but rarely in afternoon, which lead to the same video frame captured in evening enjoy the most drastic light and shadow oscillation. Therefore,

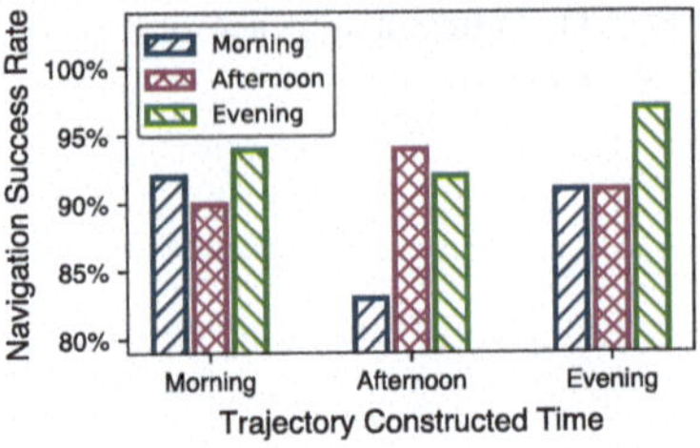

Fig. 5.14 Different time during a day

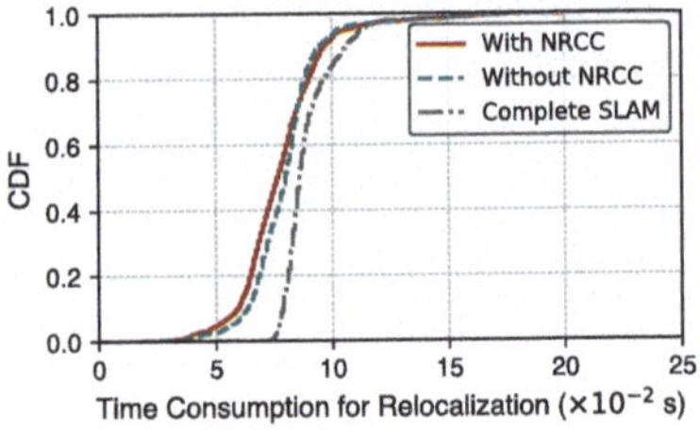

Fig. 5.15 System latency

the video frame has more ORB feature points than captured in morning and afternoon [20], and the relocalization success rate of the frame will increase.

5.5.2.4 System Latency

We recorded the time consumption of all frames in all followers' navigation experiments. As shown in Fig. 5.15, Pair-Navi reduces the average relocalization time for one frame to 76 ms. Compared with a complete SLAM system, the average delay is reduced by 17.4%; moreover, the percentage of frames using less than 100 ms for relocalization increases from 81.6 to 91%. Surprisingly, we also observe that NRCC even slightly reduces the system latency by 4 ms per frame in average. This is because NRCC shrinks the number of valid feature points involved in relocalization. The average mask process time for each frame (resized as 640×480) is 0.18 s in our system. In other words, a mask takes in charge of the NRCC of two frames that are captured in 0.2 s. In a nutshell, Pair-Navi accomplishes relocalization and navigation within the system sampling time of 0.1 s and runs fluently in real-time, with partial computation offloaded to a cloud server. As our future work, we plan to optimize to a complete standalone system on smartphones based on model compressing.

5.5.2.5 Energy Consumption

We record the energy consumption of the follower's navigation app on the smartphones. On the Huawei P10, the program ran 41 min and 48 s and consumed 203.12 mAh, while the battery level dropped from 100 to 69%. On the Lenovo Phab2pro, the program ran 44 min and 36 s and consumed 348 mAh, while the battery level dropped from 100 to 85%. Since indoor pathways from one place to another are usually less than 15-min walking distance, we consider the energy consumption of Pair-Navi is acceptable.

5.5.3 *Impact of NRCC*

To demonstrate the effect of non-rigid context culling, we first compare the relocalization failure rate with and without non-rigid context culling. Specifically, we save all the video frames captured by follower's camera and record whether each frame can be relocalized (matched to a keyframe in the trajectory map with enough feature point matches). This part of experiment is conducted on all 21 navigation trajectories mentioned above and under different time intervals. Furthermore, we have performed an extra extensive experimental validation of NRCC on two standard benchmarks for visual SLAM: TUM [17] and KITTI [18] datasets. We mainly focus on two targets: *Pose Tracking Accuracy* and *Mapping Accuracy*, and we compare the precision of ORB-SLAM with and without NRCC.

5.5.3.1 Performance in Navigation Scenario

As shown in Fig. 5.16, the relocalization failure rate increases from 4 to 27% without NRCC in the same day, 1 week later, and 2 weeks later, while it keeps low at 3%, 5%, and 9%, respectively, with NRCC. In other words, the use of NRCC significantly declines the relocalization failures by more than 65% when the time interval exceeds 2 weeks.

Furthermore, Fig. 5.17 shows the relocalization robustness at different areas. We conducted the experiment under the time interval of 2 weeks. The relocalizaiton failure rate at the office building, the gymnasium, and the shopping mall is 2%, 5%, and 10% with NRCC, compared to 4%, 15%, and 25% when without NRCC. In average, NRCC declines relocalization failures by 57%. Especially, in the gymnasium, the failure rate is decreased by 67%, which reflects remarkable improvement on robustness.

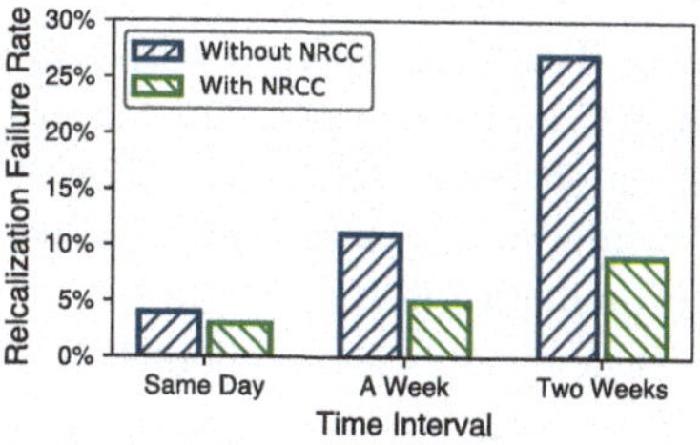

Fig. 5.16 Impact of NRCC over different time intervals

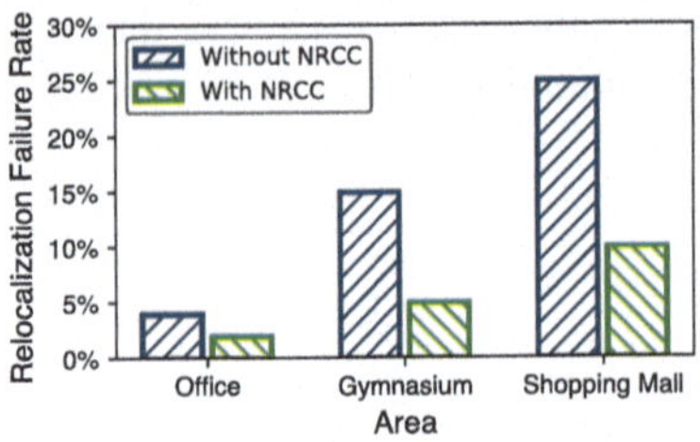

Fig. 5.17 Impact of NRCC in different areas

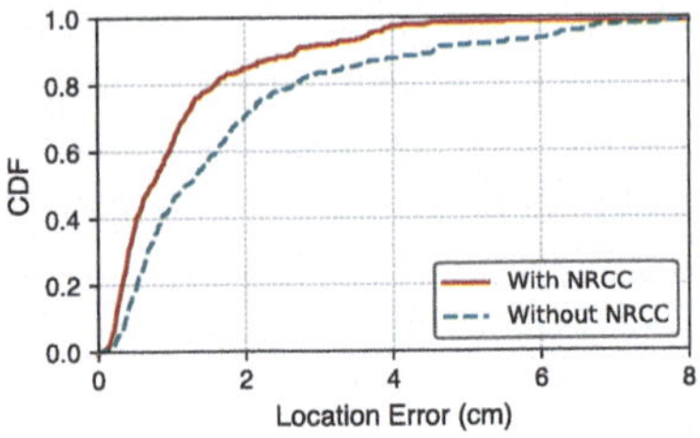

Fig. 5.18 Localization accuracy on TUM

5.5.3.2 Performance in Standard Datasets

The TUM benchmark is an excellent dataset to evaluate the accuracy of camera localization as it provides several sequences with accurate ground truth obtained with an external motion capture system. The odometry benchmark from the KITTI dataset contains 11 sequences from a car driven around a residential area with accurate ground truth from GPS and a Velodyne laser scanner. This is an exceedingly challenging dataset for monocular vision due to fast rotations and areas with lots of dynamic objects, which make data association more difficult.

We evaluate the general localization accuracy in 16 handheld sequences of the TUM RGB-Monocular benchmark. Moreover, in five sequences from the KITTI dataset, we evaluate the tracking accuracy of camera pose and precision of the constructed map.

Localization Accuracy in TUM Dataset Figure 5.18 depicts the impact of the proposed NRCC in indoor localization scenarios. As shown, the average localization of ORB-SLAM with and without NRCC is 1.24 cm and 1.97 cm, respectively. The use of NRCC significantly increases the localization precision by more than 35%.

Pose Tracking Accuracy in KITTI Dataset We further examine the impact of NRCC on camera pose tracking task in KITTI dataset. We calculate the Relative Rotation Error (RRE, an essential evaluation indicator in KITTI dataset) of the tracking result with and without NRCC. As shown in Fig. 5.19, the tracking accumulative bias is within 0.25° for 1 m length traces with NRCC, compared to 0.34 degree/m when without NRCC. In average, NRCC declines RRE by 25.6%.

Mapping Precision in KITTI Dataset Finally, we evaluate the impact of NRCC on mapping accuracy in KITTI dataset. Relative Translation Error (RTE, another fundamental evaluation indicator in KITTI dataset) is used to demonstrate the

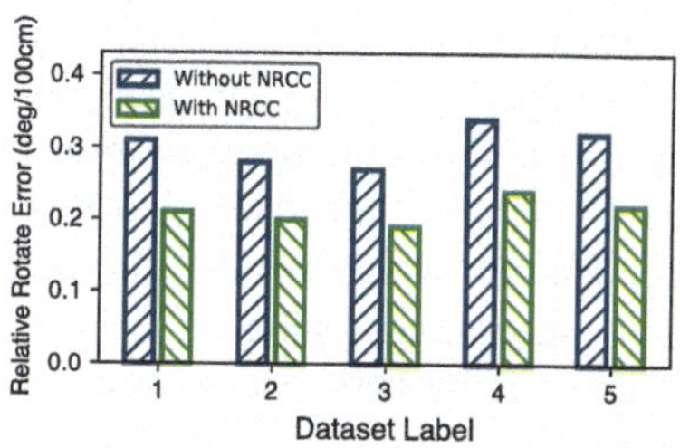

Fig. 5.19 Tracking accuracy on KITTI

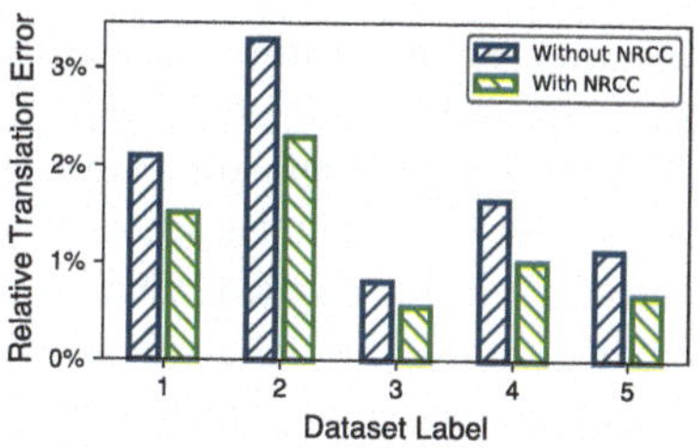

Fig. 5.20 Mapping accuracy on KITTI

mapping accuracy. As shown in Fig. 5.20, in all five frame sequences, NRCC declines RTE by 10%.

In summary, the leverage of NRCC achieves enhanced localization accuracy and mapping precision compared with original ORB-SLAM in all evaluations. The delightful performance gain owes to the robust design of NRCC which attacks environment change by ruling out the changed while maintaining the unchanged contexts in the trajectory map.

5.6 Related Works

Peer-to-Peer Indoor Navigation Traditional indoor wayfinding solutions typically necessitate a comprehensive global mapping and are reliant on infrastructure within indoor settings such as Wi-Fi [30–32], Bluetooth, and RFID [33]. These methods generate guidance based on the user's absolute position within a universal layout. Recent advances have introduced Peer-to-Peer (P2P) navigation as a viable alternative, eschewing the need for complete global mappings and absolute location pinpointing [1–3, 34–36]. Among these, *Travi-Navi* [3] and *FollowMe* [2] leverage innovative trace-driven navigation methods on smartphones, utilizing a mix of Wi-Fi, inertial measurements, and magnetic sensing to guide users without the need for an indoor floor plan. *ppNav* [1] introduces an ingenious use of Wi-Fi fingerprints, coupled with diagrammatic visual and radio feature extraction for relative location tracking. In stark contrast, Pair-Navi harnesses vision technology, offering an infrastructure-independent, more accurate, and efficient navigation solution that further facilitates various vision-based applications.

Advancements in Monocular Visual SLAM Initial explorations in visual SLAM [4] utilized a filtering approach, but later advancements [5, 37] showcased the superior accuracy of optimization-based methods [38]. Notably, *ORB-SLAM*[6] emerged as a leading monocular visual SLAM system, incorporating *DBoW2*[25] for place recognition and *g2o* [39] for optimization. Contemporary research is branching into integrating semantics for enhanced environmental adaptability [12, 13] and incorporating IMU data to address monocular SLAM's scale uncertainty [40, 41], marking significant milestones for visual SLAM's application in indoor navigation.

Instance Segmentation Evolution Instance segmentation has seen significant progress, initially propelled by the R-CNN [9] technique, which evaluated CNNs for object detection in predefined regions. This led to methods like DeepMask [42] and its successors, which learned to propose segment candidates for classification. The introduction of Mask R-CNN [11] marked a paradigm shift, offering precise mask predictions for object classification by innovatively applying an RoIAlign layer for feature extraction alignment and a two-stage process for enhanced accuracy and speed, thereby setting a new standard in instance segmentation technology.

5.7 Conclusion

In this chapter, we present Pair-Navi, a robust and real-time P2P navigation system based on visual SLAM, requiring no preinstalled infrastructure or pre-deployed localization services. We decouple conventional visual SLAM into independent modules and reassemble the necessary components into a P2P navigation system. To bolster system performance in time-varying environments, we incorporate Mask R-CNN to dynamically cull non-rigid environmental changes. We implement Pair-Navi on commodity smartphones and conduct experiments in multiple buildings over 2 weeks. Experiment results show that our system outperforms existing solutions in navigation success rate and robustness. We believe Pair-Navi takes a promising step toward practical P2P navigation. Our future works target at fusing crowdsourced trajectories to generalize navigation routes to a larger scale and make a global consistent map.

References

1. Z. Yin, C. Wu, Z. Yang, Y. Liu, Peer-to-peer indoor navigation using smartphones. IEEE J. Sel. Areas Commun. **35**(5), 1141–1153 (2017)
2. Y. Shu, K.G. Shin, T. He, J. Chen, Last-mile navigation using smartphones, in *Proceedings of ACM MobiCom* (2015)
3. Y. Zheng, G. Shen, L. Li, C. Zhao, M. Li, F. Zhao, Travi-Navi: Self-deployable indoor navigation system, in *Proceedings of the ACM MobiCom* (2014)

4. A.J. Davison, Real-time simultaneous localisation and mapping with a single camera, in *Proceedings of the IEEE ICCV* (2003)
5. G. Klein, D. Murray, Parallel tracking and mapping for small AR workspaces, in *Proceedings of the IEEE ISMAR* (2007)
6. R. Mur-Artal, J.M.M. Montiel, J.D. Tardós, ORB-SLAM: A versatile and accurate monocular SLAM system. IEEE Trans. Rob. **31**(5), 1147–1163 (2015)
7. C. Cadena, L. Carlone, H. Carrillo et al., Past, present, and future of simultaneous localization and mapping: toward the robust-perception age. IEEE Trans. Rob. **32**(6), 1309–1332 (2016)
8. R.A. Newcombe, D. Fox, S.M. Seitz, DynamicFusion: Reconstruction and tracking of non-rigid scenes in real-time, in *Proceedings of IEEE CVPR* (2015)
9. R. Girshick, J. Donahue, T. Darrell, J. Malik, Rich feature hierarchies for accurate object detection and semantic segmentation, in *Proceedings of the IEEE CVPR* (2014)
10. S. Ren, K. He, R. Girshick, J. Sun, Faster R-CNN: Towards real-time object detection with region proposal networks, in *Advances in Neural Information Processing Systems* (2015)
11. K. He, G. Gkioxari, P. Dollár, R. Girshick, Mask R-CNN, in *Proceedings of the IEEE ICCV* (2017)
12. S.L. Bowman, N. Atanasov, K. Daniilidis, G.J. Pappas, Probabilistic data association for semantic slam, in *Proceedings of the IEEE ICRA* (2017)
13. J. Civera, D. Gálvez-López, L. Riazuelo, J.D. Tardós, J. Montiel, Towards semantic slam using a monocular camera, in *Proceedings of the IEEE IROS* (2011)
14. R.F. Salas-Moreno, R.A. Newcombe, H. Strasdat, P.H. Kelly, A.J. Davison, Slam++: Simultaneous localisation and mapping at the level of objects, in *Proceedings of IEEE CVPR* (2013)
15. *ROS Kinetic Kame*. http://wiki.ros.org/kinetic
16. *Ros android*. http://wiki.ros.org/android
17. J. Sturm, N. Engelhard, F. Endres, W. Burgard, D. Cremers, A benchmark for the evaluation of RGB-D slam systems, in *Proceedings of the IEEE IROS* (2012)
18. A. Geiger, P. Lenz, R. Urtasun, Are we ready for autonomous driving? The KITTI vision benchmark suite, in *Proceedings of the IEEE CVPR* (2012)
19. R. Hartley, A. Zisserman, *Multiple View Geometry in Computer Vision* (Cambridge University Press, Cambridge, 2003)
20. E. Rublee, V. Rabaud, K. Konolige, G. Bradski, ORB: An efficient alternative to sift or surf, in *Proceedings of the 2011 International Conference on Computer Vision (ICCV)* (IEEE, Piscataway, 2011), pp. 2564–2571
21. P. C. Ng, S. Henikoff, SIFT: Predicting amino acid changes that affect protein function. Nucl. Acids Res. **31**(13), 3812–3814 (2003)
22. H. Bay, T. Tuytelaars, L. Van Gool, SURF: Speeded up robust features, in *European Conference on Computer Vision* (Springer, Berlin, 2006), pp. 404–417
23. J.J. Craig, *Introduction to Robotics: Mechanics and Control*, vol. 3 (Pearson/Prentice Hall, Upper Saddle River, 2005)
24. V. Lepetit, F. Moreno-Noguer, P. Fua, EPnP: An accurate O(n) solution to the PnP problem. Int. J. Comput. Vision **81**(2), 155 (2009)
25. D. Gálvez-López, J.D. Tardos, Bags of binary words for fast place recognition in image sequences. IEEE Trans. Rob. **28**(5), 1188–1197 (2012)
26. T.-Y. Lin, M. Maire, S. Belongie et al., Microsoft coco: Common objects in context, in *European Conference on Computer Vision* (Springer, Berlin, 2014)
27. J. Redmon, S. Divvala, R. Girshick, A. Farhadi, You only look once: Unified, real-time object detection, in *Proceedings of the IEEE CVPR* (2016)
28. A. Rai, K.K. Chintalapudi, V.N. Padmanabhan, R. Sen, Zee: Zero-effort crowdsourcing for indoor localization, in *Proceedings of the ACM MobiCom* (2012)
29. Y. Shu, Y. Huang, J. Zhang et al., Gradient-based fingerprinting for indoor localization and tracking. IEEE Trans. Ind. Electron. **63**(4), 2424–2433 (2016)
30. Z. Yang, C. Wu, Y. Liu, Locating in fingerprint space: Wireless indoor localization with little human intervention, in *Proceedings of the ACM MobiCom* (2012)

31. K. Qian, C. Wu, Y. Zhang, G. Zhang, Z. Yang, Y. Liu, Widar2.0: Passive human tracking with a single Wi-Fi link, in *Proceedings of ACM MobiSys* (2018)
32. Z. Yang, Z. Zhou, Y. Liu, From RSSI to CSI: indoor localization via channel response. ACM Comput. Surv. **46**(2), 25 (2013)
33. L. Shangguan, Z. Yang, A.X. Liu, Z. Zhou, Y. Liu, STPP: spatial-temporal phase profiling-based method for relative RFID tag localization. IEEE/ACM Trans. Netw. **25**(1), 596–609 (2017)
34. T.H. Riehle, S.M. Anderson, P.A. Lichter et al., Indoor magnetic navigation for the blind, in *Proceedings of IEEE EMBC* (2012)
35. W. Storms, J. Shockley, J. Raquet, Magnetic field navigation in an indoor environment, in *Proceedings of IEEE UPINLBS* (2010)
36. I. Constandache, X. Bao, M. Azizyan, R.R. Choudhury, Did you see bob?: Human localization using mobile phones, in *Proceedings of ACM MobiCom* (2010)
37. E. Mouragnon, M. Lhuillier, M. Dhome, F. Dekeyser, P. Sayd, Real time localization and 3d reconstruction, in *Proceedings of the IEEE CVPR* (2006)
38. H. Strasdat, J.M. Montiel, A.J. Davison, Visual slam: why filter? Image Vision Comput. **30**(2), 65–77 (2012)
39. R. Kümmerle, G. Grisetti, H. Strasdat, K. Konolige, W. Burgard, G^2o: A general framework for graph optimization, in *Proceedings of the IEEE ICRA* (2011)
40. C. Forster, L. Carlone, F. Dellaert, D. Scaramuzza, On-manifold preintegration for real-time visual–inertial odometry. IEEE Trans. Rob. **33**(1), 1–21 (2017)
41. R. Mur-Artal, J.D. Tardós, Visual-inertial monocular slam with map reuse. IEEE Rob. Autom. Lett. **2**(2), 796–803 (2017)
42. P.O. Pinheiro, R. Collobert, P. Dollár, Learning to segment object candidates, in *Advances in Neural Information Processing Systems* (2015)

Chapter 6
Large-Scale Crowdsourced Mapping with EdgeSLAM

Abstract Indoor navigation complexities stem from the extensive setup and reliance on existing infrastructures. Recent advancements have explored P2P navigation to simplify setup processes and visual techniques to forgo infrastructure dependency. Yet, challenges persist: P2P navigation often grapples with path scarcity, and visual techniques on smartphones face issues such as scale uncertainty, directionality limitations, and high computational demands. Addressing these hurdles, our innovation introduces a visual P2P navigation system enhanced by an extensive crowdsourcing framework, effectively mitigating these issues. Through theoretical insights and empirical evidence, our approach demonstrates the potential of P2P navigation—historically constrained by limited path availability—to flourish with an enriched crowdsourced map.

We leverage IMU data to overcome scale discrepancies and develop a novel navigational approach to counteract the issue of directionality. Our proposed mobile-edge computing framework facilitates swift navigation (30 FPS with a 100 ms end-to-end latency) while significantly reducing smartphone workload (extending battery life by 35% for a continuous 2 h and 35 min of use), ensuring both precise localization and map accuracy. Our testing confirms a navigation success rate of 100% and a spatial deviation less than 3.2 meters, surpassing previous methodologies in performance.

6.1 Introduction

Indoor navigation showcases potential across various domains, such as expansive retail environments and logistics operations. Traditional approaches to indoor navigation involve positioning individuals within a comprehensive map to guide them [1]. Implementations utilizing Wi-Fi [2, 3], RFID [4, 5], and acoustic signals [6] have proven effective. Nonetheless, these methods necessitate the creation of indoor radio maps, a process that demands significant labor and time. Additionally, their effectiveness hinges on the availability of existing indoor infrastructure or the need for specific installations, limiting their practicality.

J. Xu et al., *Edge Assisted Mobile Visual SLAM*,
https://doi.org/10.1007/978-981-97-3573-0_6

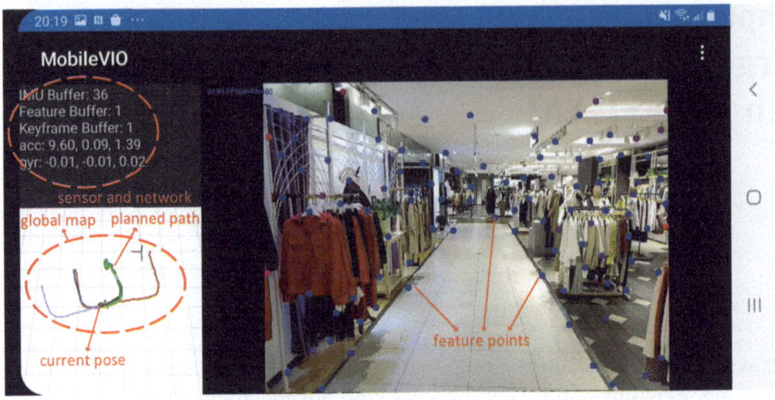

Fig. 6.1 User interface of the mobile side. On the left bottom, the current pose of the follower is shown by a small red pyramid and the planned path in green. On the left top lies the sensor information and network status. On the right is the latest image. Visual navigation is user-friendly because the follower can have a good understanding of his/her position and the path

Peer-to-peer (P2P) navigation emerges as a novel paradigm, eliminating the extensive requirements for site surveys [7]. This approach involves an initial user, termed the leader, documenting path details, including sensor data at pivotal points. Subsequently, this path information aids a subsequent user, referred to as the follower, aiming to traverse the same route from origin to destination. The follower replicates the leader's documented cues to navigate the pathway. Empirical studies utilizing wireless signals and inertial sensors illustrate the efficacy of this P2P approach in guiding users successfully [8, 9].

On the other hand, in recent years, the rosy field of computer vision sheds light on eliminating indoor infrastructure. Based on multi-view geometry, visual simultaneous localization and mapping (SLAM) takes video frames as input and outputs more accurate trajectories and maps than previous methods. Additionally, another advantage of visual navigation lies in more user-friendliness, as images are richer in semantics than wireless signal readings such that users can directly understand the paths they are taking (see Fig. 6.1 for an example). Works like *Pair-Navi* [10] have incorporated the idea of SLAM into peer-to-peer navigation. It generates a trajectory and a local map for the video of each leader, and it can navigate effectively, run in real-time, and keep robustness for a long time. However, although *Pair-Navi* works fine with motivated leaders (it encourages shop owners to build trajectories from the front gate of the mall to his/her shop), it has a relatively small path archive when it comes to large indoor scenarios with multifarious combinations of origins and destinations, such as paths between shops. This deficiency of paths, vitiating the practicability of P2P navigation, is attributed to the nature of the P2P style.

In this chapter, we seek to push the limit of visual P2P navigation systems toward better applicability. Regarding path deficiency, intuitively, if we can merge local maps into a consistent global map and generate paths by splicing them, we can obtain a more comprehensive path coverage. *FollowUs* [11] first embodied the idea of splicing local maps in magnetic navigation. However, in visual navigation, this idea entails several challenges.

First, monocular visual SLAM suffers from scale ambiguity, the meaning of which is twofold: (1) Different trajectories have different unknown scales, thus impossible to splice together directly, and (2) within one trajectory, the scale can drift if the trajectory is very long. Despite this major shortcoming, we have to use monocular navigation because most smartphones have only one camera, and even those equipped with dual cameras have inconsistent lens types and short baselines that do not suffice stereo visual SLAM. **Second**, visual SLAM has the unidirectional problem, which means the trajectories and map construction can only be reused in one direction, since the same spot has disparate scenes in two opposite directions. Therefore, even if we construct a global map by splicing local maps, it is impossible to use them directly.

To tackle the issue of scale ambiguity, we leverage the IMU sensor, known for providing measurements in meters, merging visual and inertial data to create visual-inertial odometry (VIO). This approach delivers outcomes akin to visual SLAM but with trajectories aligned to a consistent metric scale. To address the challenge of unidirectional navigation, we have conceptualized a strategy that enables navigation within a global map accommodating both directions of travel. Yet, this solution introduces a **third** challenge. Given that VIO depends on optimizing high-dimensional matrices, executing it on smartphones in real-time without compromising trajectory and map precision is challenging. High accuracy is crucial for crafting a coherent global map, and the extensive battery consumption for optimization further exacerbates the problem. Consequently, there is a pressing need to alleviate the computational demand on mobile devices. Our solution employs a mobile-edge VIO framework, strategically partitioning the VIO process across mobile devices and edge servers. This setup optimizes data exchanges to minimize both the computational load on mobile devices and the bandwidth for mobile-edge communication, ensuring both precision and swift responsiveness. Introducing Crowd-Navi (Global Map P2P Navigation, with the interface depicted in Fig. 6.1), this system enhances peer-to-peer navigation by incorporating a global map through an extensive crowdsourcing approach. This innovation allows for a plethora of navigational paths, significantly improving the system's utility and applicability.

This chapter introduces the following key advancements:

- We elevate the viability of visual P2P navigation by enhancing path availability through a detailed crowdsourcing approach. This method sees path coverage exponentially increase at a rate of $O(m!)$ (m representing the count of individual paths) during the initial phase, marking a first in both theoretical and experimental validation of such path coverage benefits.
- We liberate crowdsourced visual P2P navigation from constraints such as scale ambiguity and directional limitations. The integration of visual and inertial data

overcomes scale ambiguity, while a novel navigation approach addresses the directional challenge.
- We facilitate precise, real-time navigation on mobile devices with reduced computational demand, courtesy of an innovative edge-supported visual-inertial odometry framework.

Experiments demonstrate that besides comprehensive path coverage, our system can navigate with a navigation success rate of 100%. The spatial offset of our system is less than 3.2 m in environments like shopping malls, supermarkets, teaching buildings, and office buildings, better than existing works. Additionally, our mobile-edge architecture guarantees the lightweight burden of the smartphone (35% battery life for 2 h 35 min) and real-time operation (30 FPS, as specified in [12, 13] and 100 ms end-to-end delay).

The rest of this chapter is organized as follows: We introduce the fundamentals of VIO in Sect. 6.2, we detail the navigation workflow in Sect. 6.3 and the mobile-edge architecture in Sect. 6.4, and after that, we discuss the experiments and conclude the chapter.

6.2 Preliminaries

6.2.1 *Monocular Visual-Inertial Odometry*

Monocular VIO takes video frames and IMU readings as input and outputs the real-time pose of the camera and a local map of the 3D point landmarks (corresponding to 2D feature points in the images).

6.2.1.1 Visual Model and IMU Model

A VIO system contains the measurements from two models: the visual model and the IMU model. Its central idea is to optimize the camera poses and landmark positions so that the difference between the theoretical observations and the practical measurements is minimized.

In the visual model, the theoretical observation is the 2D pixel coordinate of a feature point, which represents a 3D point landmark reprojected onto the camera plane. The 2D pixel coordinate is computed by $p = h(t_{wc}, R_{wc}, P_w)$, where t_{wc}, R_{wc} comprise the pose (position and rotation described in a 3D vector and a third-order orthogonal matrix) of a camera, P_w is the 3D position of the point landmark in the world coordinate, and $h(\cdot)$ is the observation function outputting a 2D vector. On the other hand, the practical measurement is the reading of the feature point's pixel coordinate in the image, which is also a 2D vector. By optimizing t_{wc}, R_{wc}, P_w, we can minimize the difference between the measured p and the theoretical $h(t, R, P_w)$.

The IMU model is about a notion called "preintegration" [14], which reflects the relative motion increment between two poses and is computed from IMU readings sampled between the time interval of the two images. Coming exclusively from IMU measurements, preintegration can be regarded as a practical measurement, though having been processed. Besides, if we have the poses of a camera at two adjacent images, t_i, R_i, t_{i+1}, R_{i+1}, we can also compute the relative motion increment from these poses, which is the theoretical observation of the IMU model. By optimizing t_i, R_i, t_{i+1}, R_{i+1}, we can minimize the difference between the relative motion increment computed by the poses (theoretical observation) and IMU preintegration (practical measurement).

6.2.1.2 Sliding Window Tracking and Optimization

After an initialization step [15, 16], the system starts to maintain a sliding window and optimize the most recent states and point landmarks within it. When a new video frame comes, all the IMU readings between this frame and the last frame are calculated as a preintegration value. Then, combining all the information in the sliding window, i.e., the visual models, IMU preintegrations, and some prior information, an optimization problem is formed as

$$\min_{X}\left\{ \left\|r_p(X)\right\|^2 + \sum_{k\in B} \left\|r_B(z^{b_k}_{b_{k+1}}, X)\right\|^2 \tag{6.1}$$

$$P^{b_k}_{b_{k-1}} + \sum_{(l,j)\in C} \rho(\left\|r_C(z^{c_j}_l, X)\right\|)^2_{P^{c_j}_l}) \right\}, \tag{6.2}$$

where X is all the states and point landmarks inside the sliding window, B is the set of all the preintegrations in the sliding window, and C is the set of all the camera observations where (l, j) is the j-th camera observing the l-th landmark. The IMU observations are written as $z^{b_k}_{b_{k+1}}$, where b_k stands for "body pose of k-th frame," and the visual observations $z^{c_j}_l$, where c_j stands for "camera pose of j-th frame." The residual of the IMU model, the visual model, and the prior term are written as r_B, r_C, and $r_p(\cdot)$. $\rho(\cdot)$ is a robust kernel function to eliminate the effect of outliers, i.e., the mismatched feature points. The additive items are all squared norms where $P^{b_k}_{b_{k-1}}$ and $P^{c_j}_l$ are Mahalanobis weights. After intensive computation, this optimization problem outputs a series of camera poses and the positions of the point landmarks that minimizes both visual and IMU models simultaneously. Then, when new video frame comes, the sliding window moves forward, marginalizing out old states and solving a new optimization problem. Repeating the above steps, the system keeps an updated estimation of the latest states and the point landmarks.

6.2.1.3 Keyframe Selection

For every video frame, if it observes a large parallax with its predecessor, we take it as a keyframe. Collecting all the keyframes and all the point landmarks, we acquire a leader's trajectory, as well as the corresponding local map, and in the following of this chapter, we call the combination of them a "sequence" with an inborn direction in which it is collected. A sequence is denoted by two strings, its origin name and its destination name. This way, its direction is implicit in the two names.

6.2.2 Relocalization and Sequence Splicing

Suppose we already have one sequence contributed by a leader, which is regarded as the primitive global map.

6.2.2.1 Relocalization

When a second leader or a follower comes, our system compares every video frame of his/hers with the keyframes in the previous sequence and finds out the most similar keyframe as a relocalization candidate keyframe. If the number of feature point matches between this candidate keyframe and the current frame surpasses a threshold and if the candidate keyframe and the current frame passes a fundamental matrix check [17], then the candidate is regarded as a relocalization keyframe. This procedure is called relocalization. When more sequences are in the global map, we search in all the previous sequences for a relocalization keyframe.

6.2.2.2 Sequence Splicing

Since the new sequence shares the same metric scale with the global map, to splice them together, we simply need to calculate a relative pose transformation from the new sequence to the global map.[1] As there might be consecutive relocalizations occurring in the sliding window, we consider all of them and solve a perspective-n-point problem [18]

$$\min_{R,t} \sum_{L \in Rel} \sum_{(l,v) \in L} \left\| r_C(z_l^v, RP_l^v + t) \right\|^2, \tag{6.3}$$

[1] As opposed to pure monocular odometry sequences, each of which has unknown scale such that they cannot be spliced.

where R, t is the rotation and the translation of this transformation, Rel is the set of all relocalization frames, (l, v) means the l-th feature observed in the v-th relocalization frame, and P_l^v is the position of the point landmark associated with feature point (l, v) in the coordinate of the global map, so $(RP_l^v + t)$ means the position of the point landmark in the coordinate of the new sequence. Again, the visual residual function r_C is to measure the difference between the pixel coordinate z_l^v and the reprojection of the 3D point landmark. With the solved pose transformation, we can transform the new sequence into the global map, merging them into a whole. Unfortunately, some angle error may cause the merged global map to be wrongly shaped. So the merged map will further go through a 4-DoF (degree-of-freedom) smoothing [19].

6.3 System Design for EdgeSLAM-Based Crowdsourced Mapping

In this section, we show in detail how a global map is constructed and how we navigate a follower with our crowdsourcing strategy. Then we demonstrate the algorithms for calculating navigation paths.

6.3.1 Global Map Construction

We consider three major situations about how sequences should be crowdsourced to form a global map, as illustrated in Fig. 6.2. These situations are only examples of the possibility of merging, regardless of the shapes of the sequences.

Situation 1. Leader 1 contributes seq1 from the gate of the mall to shop A, which is regarded as the global map, and leader 2 contributes seq2 from the gate to shop B. In this case, since seq1 and seq2 contain a common segment (Gate->C) with the identical direction, relocalization happens and seq2 is merged into the global map (i.e. seq1).

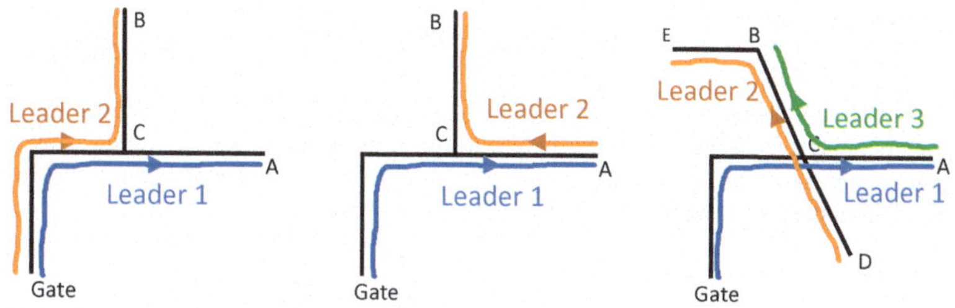

Fig. 6.2 Three situations of the leader

Situation 2. Leader 1 remains the same. Leader 2 goes from A to B. Even though two sequences share the same passage (AC), splicing is precluded because they are in mutually inverse directions, hence no relocalization. To splice more sequences, our system requires the leaders to face the camera toward the shopfronts at the beginning and in the end. In this specific case, Leader 2 going from the gate to shop A should first face his camera toward the gate and then begin walking and shooting video. When he arrives at A, he turns the camera toward the shopfront of A and finishes the leading procedure. Similarly, Leader 2 should first face his camera toward A. Owing to this requirement, the paths will have co-vision and can be spliced together.

Situation 3. Leader 1 remains the same, and leader 2 contributes seq2 from shop D to shop B. Since no relocalization occurs, they cannot be merged. In our design, if a sequence has no relocalization with any of the existing sequences, we simply store this sequence, and it may be merged into the global map through future sequences. When Leader 3 contributes seq3 from A to B, it first detects relocalization with seq1 at A and merges itself into the global map. Then, at C, seq3 relocalizes with seq2. Since seq3 is already in the global map, seq2 is merged in the global map as well.

In a nutshell, the three situations suggest that no matter what shape a path is in, it can contribute to the global map if (1) it has relocalization with previous paths, (2) it has the same origin or destination with previous paths, and (3) it can be connected with previous paths through future paths. A leader's workflow is demonstrated in Fig. 6.3. An edge server generates a corresponding sequence according to the leader's moving path. Note that sequences are spliced through relocalizations, and later when we discuss the sequence shift in Sect. 6.3.3, the shifting spots mean the relocalization keyframes.

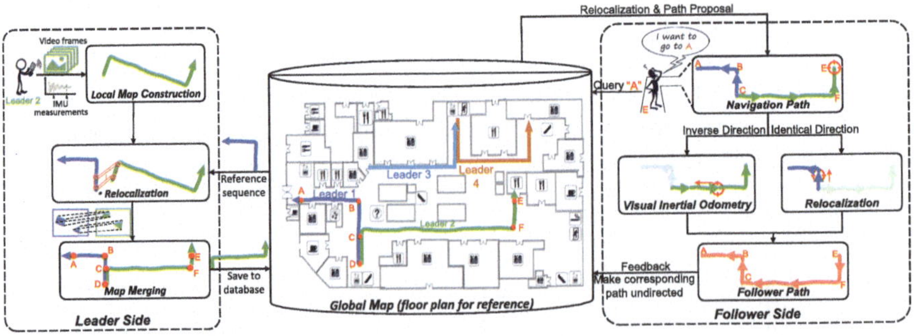

Fig. 6.3 Workflow of a leader and a follower. Leaders and followers all interact with the global map. Leaders build a local map and contribute it to the global map; followers use the global map to navigate, either by VIO in inverse directions or by relocalization in identical directions, and can also contribute to the global map

6.3.2 *Navigation in Both Directions*

For each video frame of the follower, *Pair-Navi* mainly carries out relocalization to localize the follower in the leader's trajectory. However, in Crowd-Navi, this strategy is not suitable because we need to support navigating in an inverse direction. If the follower's camera is facing toward the opposite direction of the sequence, no relocalization will happen. Therefore, we propose a VIO-based navigation method for followers.

6.3.2.1 A Typical Workflow of Navigation

Suppose a follower is going from E to A (see Fig. 6.3). He first selects the name of A from the known places of interest and queries it to the global map. Our system first instructs him to relocalize in the global map. Then he travels on the path E->F->C in an inverse direction with the original path of Leader 2.This is when our system runs a full VIO to track him and compute navigation paths accordingly. In practice, as VIO is in nature a dead-reckoning method, it gradually accumulates drifting error; however, if a relocalization is detected later, the location of the follower can be bound to the global map, resetting the drift to zero at that spot. Accordingly, a cost term of relocalization should be added to the sliding window in Eq. (6.2):

$$\min_{X}\left\{ \left\|r_P(X)\right\|^2 + \sum_{k\in B}\left\|r_B\left(z_{b_{k+1}}^{b_k}, X\right)\right\|_{P_{b_{k-1}}^{b_k}}^2 \right.$$
$$\left. + \sum_{(l,j)\in C}\rho\left(\left\|r_C\left(z_l^{c_j}, X\right)\right\|_{P_l^{c_j}}^2\right) + \sum_{L_i\in R}\sum_{(l,v)\in L_i}\rho\left(\left\|r_C\left(z_l^{v}, X\right)\right\|_{P_l^{c_v}}^2\right)\right\}, \tag{6.4}$$

where Rel is the set of all the relocalization frames in the sliding window, L_i is the set of matched features in one relocalization frame, (l, v) is l-th feature observed in the frame v, and $P_l^{c_v}$ is another Mahalanobis weight.

When the follower turns right at C, his smartphone is relocalized because now the camera is facing toward the direction identical to the sequence. After this relocalization, the drifting error of this frame is corrected to zero and begins to accumulate from zero again. In a large global map, it is often the case that a follower will meet several relocalizations, so the drifting error of VIO is bounded and the system works fine. But although relocalization can correct drift, the performance of navigation in inverse direction is still crucial, which we will demonstrate in Sect. 6.5.1.

In short, the navigation strategy is to carry out VIO the whole time, and whenever a relocalization happens, it corrects drift. After the navigation, the follower's path is fed back to the global map, just like a leader. In this way, followers help enrich the global map.

6.3.2.2 A Perspective of Graph Theory

From a perspective of graph theory, since the sequences are innately directed, the navigation takes place in a directed graph, where each path is supposed to be unidirectional. That said, we manage to navigate even in an inverse direction with the slight disadvantage of VIO's drift. Furthermore, once the follower finishes walking in an inverse direction of a passage and feeds back his sequence to the global map, this passage becomes undirected. In this way, the global map becomes more accurate for navigation, because we have more opportunities to relocalize rather than have to carry out VIO for a long time.

6.3.3 Path Planning

To plan navigation paths, we first need to find out if the destination is reachable from the follower's current position. We use a disjoint-set data structure [20], as well as a connected graph to manage the sequences. In this way, Crowd-Navi can find if two sequences are connected in $O(\alpha(n))$ time complexity, where $\alpha(\cdot)$ is the Ackermann function. When a follower chooses his destination name, Crowd-Navi finds a sequence containing this name and regards it as the destination sequence; after Crowd-Navi relocalizes the follower in the global map, the origin sequence is also known. Then Crowd-Navi carries out a disjoint-set search to find if the destination sequence is reachable from the origin sequence. If the destination sequence is reachable which means the destination spot is reachable (we can navigate a follower in both ways), then Crowd-Navi can plan a path.

6.3.3.1 Peer-to-Peer Path Planning

To plan paths in a P2P manner, our system first searches in the connected graph of sequences for a sequence shift order, i.e., the series of sequences a follower should travel. With this sequence shift order and the relocalization keyframes as the shifting spots, Crowd-Navi puts the needed keyframes in order into a navigation path, which is illustrated in Fig. 6.4a.

6.3.3.2 Keyframe-Level Path Planning

Although it generates valid navigation paths, P2P path planning is not always efficient. For example, if two sequences are merged at the ends, then the follower may need to detour (Fig. 6.4a). Therefore, we add another path planning scheme that tries to derive more efficient paths on a keyframe level. We maintain a K-D tree of the positions of all the keyframes. If the P2P navigation path is longer than three times the line distance between the follower's current location and the destination

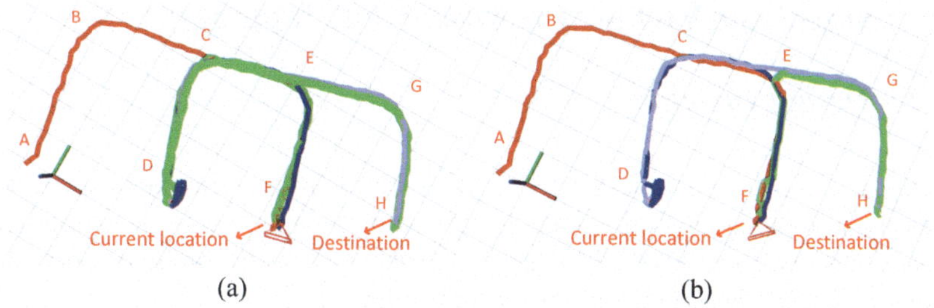

Fig. 6.4 Path proposal methods: (**a**) peer-to-peer path proposal and (**b**) active path proposal. The planned paths are shown in green lines in both cases. In peer-to-peer path proposal, a follower going from F to H needs to travel in the order "F-E-C-D-C-E-G-H," which is a huge detour, while in keyframe-level path proposal, the follower takes the nearest path "F-E-G-H"

keyframe, we search in the K-D tree on a keyframe level to see if there exists a shorter path. In the example of Fig. 6.4b, the keyframe-level path planning yields a shorter path, while the peer-to-peer path proposal tells the follower to take a detour.

6.3.4 Comprehensiveness of Path Coverage by Crowdsourcing

In a pure P2P navigation system, the follower can be navigated only if a leader has walked in the same way and uploaded the sequence. In Crowd-Navi, the requirement is relaxed. This is based on the property that two POIs are mutually reachable if there exists a series of intersections of sequences, regardless of the directions.

Suppose now we already have a global map, and one wishes to go to Macdonald's from Nike. Currently, there is no leader's sequence directly connecting Macdonald's and Nike, but there are two intersecting sequences each having Macdonald's and Nike as its origin or destination. With our path planning scheme, we can plan a path from Macdonald's to Nike with these two sequences. Furthermore, we can even use more than two sequences to propose a path as long as the Macdonald's and Nike are connected through a series of intersecting sequences.

In Appendix section "Appendix: Path Coverage Model", we first derive a constant lower bound for the probability of the intersecting of two sequences. Then we use the lower bound to show that the expectation of the quantity of paths connecting two arbitrary POIs grows at the speed of $O(m!)$, where m is the amount of leader's sequences and is far less than the number of POIs. This indicates that such crowdsourcing process quickly adds up the quantity of paths connecting POIs, which is priceless when the amount of leaders' sequences is low.

6.4 Large-Scale VIO for Crowdsourced Mapping

6.4.1 Design Insights

The main idea of our mobile-edge-structured VIO is to perform data capture and lightweight preprocessing on mobile devices while running heavy optimization on edge servers. We also carefully control the rate of data transmission to reduce computational overhead and minimize bandwidth usage.

We modify and extend *VINS-Mono* [21] for our system. The system architecture is shown in Fig. 6.5. It contains a vision front-end, a feature tracker, which takes raw video frames as input, extracts feature points and performs optical flow to track them. Meanwhile, the IMU preintegration module takes IMU readings and computes preintegrations between each consecutive pair of video frames. Both visual and IMU information are fed into a sliding window module, which optimizes the current pose of the smartphone, as well as the 3D positions of the feature points, keeping track of the latest position of a leader or a follower. Finally, a sequence manager module takes charge of the global map and the current sequence, detects whether there is a relocalization of the current sequence and global map, and merges them if possible.

6.4.1.1 The Edge Side

Though *VINS-Mono* provides high accuracy for pose estimation and local map construction, it heavily depends on computation-intensive optimization and is thus difficult to run on ordinary smartphones of limited computational ability and battery power. Furthermore, it also has a relatively high demand for storage, because to detect relocalization, the sequence manager first needs to construct a keyframe database. It is easy to think of curtailing the range of the sliding window and reduce the amount of keyframe in the database. While this may help to mitigate the burden on smartphones, it can take a toll on localization and mapping accuracy, which is the worst for our system because to build a globally consistent map, all local maps need

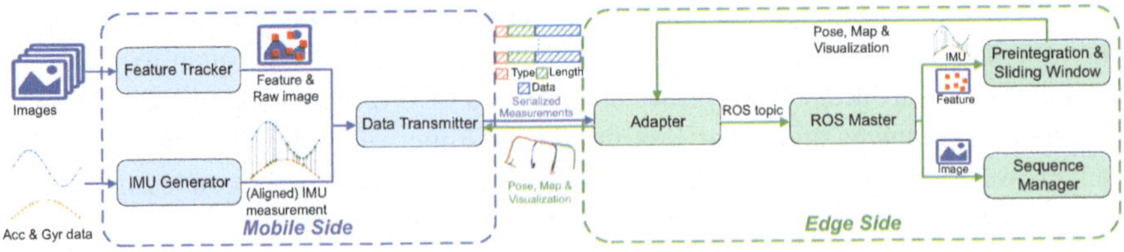

Fig. 6.5 Architecture of edge-assisted VIO. We show the mobile side on the left and the edge side on the right. The mobile side captures images and IMU and sends them to the edge side, while the edge side processes the information, manages the sequences, and sends navigation information back to the mobile side

to be as accurate as possible. Thus, to truly attenuate the burden of smartphones without losing accuracy, we put the sliding window, sequence manager modules on the edge side.

6.4.1.2 The Mobile Side

At the first glance, a lightweight mobile client should solely collect images and IMU data and send them to an edge server. However, such a solution needs a considerable amount of bandwidth due to frequent transmission of image data. For a VIO to be able to track, the frame rate should be at least 10 FPS, which adds up to a prohibitive bandwidth of 6.1 MB/s. Luckily, the sliding window only requires a set of feature points as visual measurements. With a camera taking 640×480-resolution images at 30 FPS (for better tracking and mapping accuracy under motion clutter) and each frame containing at most 100 feature points, the total bandwidth for transmitting feature points is 180 KB/s, outperforming 6.1 MB/s at 10 FPS for raw image transmission. Therefore, we extract the feature points on the mobile side and transmit them to the edge side. The whole procedure on the mobile side works in the following manner. First, we extract feature points for every frame; after that, we perform optical flow tracking to match the points to those in the previous frame and further perform a culling procedure using RANSAC with fundamental matrix model [17] (we will refer to this as "fundamental culling"), which winnows out the badly tracked points that plague the accuracy of VIO. Then the feature points are transmitted to the edge side. After that, we extract points again, to complement the total number of points for doing optical flow with the next frame (we will refer to this as "complementary extracting"). Nevertheless, though feature points are sufficient for the sliding window such that raw images are not needed here, we still need raw images for relocalization, so we also send raw images back to the edge server at a low frequency, not to congest the bandwidth.

As for the last module, the IMU preintegration module, we notice that although IMUs produce data at a rather high rate (100–300 Hz), each entry contains only several bytes of data, which means they account for only a minor portion of the total bandwidth. Also, the raw IMU readings are needed in the sliding window optimization: Because the preintegration should be recomputed after each optimization iteration, the IMU readings are repeatedly used on the edge server. Therefore, we transmit directly back to the edge the raw IMU readings, which is comprised of accelerometer and gyroscope readings firstly aligned through an IMU generator.

6.4.1.3 Pushing the Limit

With the basic architecture proposed above, we can further push the limit of the system by some more detailed tailoring strategies. We notice that as pedestrians usually move slowly (about 1.3 m/s), so considering image rate ($\geq$30 FPS), the parallax between consecutive frames is small. Therefore, we do not need to transmit back the feature points in a frame-by-frame manner but can rather do it intermittently. Regarding the frames that we do not transmit back to the edge server, we simply continue the optical flow tracking for them on the smartphone and do not need to perform fundamental culling or complementary extracting. This frugality of transmission also brings down the frequency of fundamental culling, reducing image preprocessing overhead; however, the reduction of fundamental culling can also impair tracking accuracy due to poorer feature quality. We measure the influence of feature transmission rate on tracking accuracy, image processing time, and bandwidth usage, and the results are shown in Sect. 6.5, according to which we give the choices of the feature transmission rate in different scenarios.

6.4.2 Implementation

We implement our mobile side on Android platform. The mobile consists of three main threads: a feature tracker, an IMU generator, and a data transmitter. The feature tracker extracts feature points from images and controls the transmission rate for feature points (for VIO tracking) and raw images (for relocalization). The IMU generator aligns measurements from different sensors. Since Android provides accelerometer and gyroscope measurements separately, they usually have different timestamps. For each measurement entry, the IMU generator finds its temporal neighbor and performs interpolation to generate corresponding measurements. Another problem is that some devices have different clock bases for IMU and camera (those not supported by Google ARCore certification), so we also synchronize the two clock bases by comparing them with system time for these devices. Lastly, the data transmission thread serializes each data entry into a lightweight "type-length-data" format and sent it to the edge side through a single socket.

On the edge side, we use ROS platform [22]. An adapter first deserializes the data from the mobile side and sends each type of data (feature, IMU, and raw image) through respective ROS topics. Then the preintegration module, the sliding window, and the sequence manager receive the ROS topics and carry out their functions.

6.5 Evaluation for EdgeSLAM-Based Crowdsourced Mapping

In this section, we evaluate the performance of our system both in navigation and in edge-assisted communication.

We implement our system on ROS platform. The VIO module, modified and extended from *VINS-Mono* [21], uses Ceres 1.14.0 for optimization, Eigen 3.3.4 for matrix operation, and OpenCV 3.4.2 and OpenCV Android 3.4.2 for image processing on edge server and smartphone, respectively.

6.5.1 *Navigation Performance*

6.5.1.1 Experiment Settings

We asked four volunteers to act as leaders and followers in turn, whose heights varied from 172 cm to 189 cm and whose behaviors of holding phones were varied, some holding steadily and others shakily.

The mobile phones we used were Samsung Galaxy S10+ and Huawei P10, and the edge server is Dell XPS 9560 with i7-7700HQ CPU of 2.8 GHz main frequency and 16 GB RAM, installed with Ubuntu 16.04 operating system and ROS Kinetic.

We selected a shopping mall, a teaching building, a supermarket, and an office building as our experiment venue, and the area of which is about 2000 m^2, 1000 m^2, 500 m^2, and 400 m^2, respectively.

6.5.1.2 Path Coverage

Examples of the final global maps are shown in Fig. 6.6. We can see all the places are connected and mutually reachable in the global map, thus validating a far more comprehensive path coverage. Specifically, inside a large room with four POIs and three spliced sequences, the total path number is 12 compared to three sequences (Fig. 6.6d); in a teaching building, where there are eight POIs on the second floor and three POIs on the first floor, our system generates 110 available paths, much larger than 13, the number of sequences. Thus, the path coverage grows considerably fast, making Crowd-Navi much more applicable than the pure-P2P visual navigation system *Pair-Navi*.

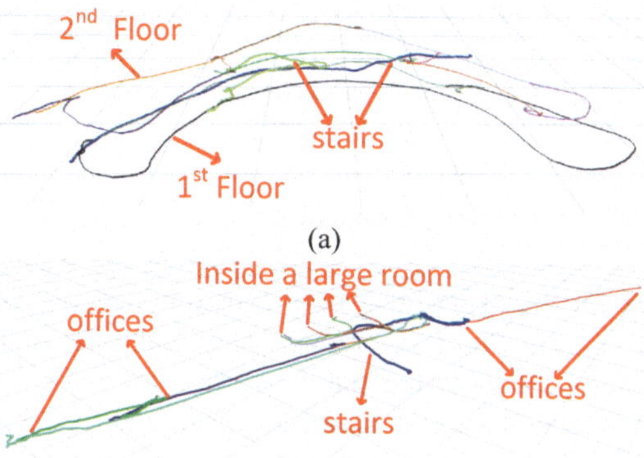

Fig. 6.6 Two global maps created in our experiment in two indoor scenarios: (**a**) two floors of a teaching building and (**b**) one floor of an office building

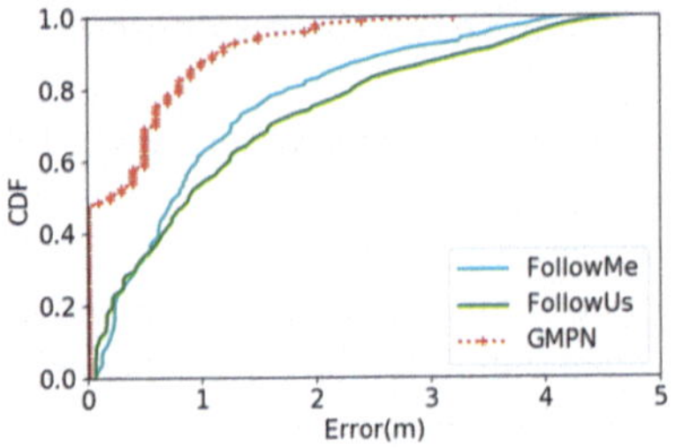

Fig. 6.7 Spatial offset comparison between Crowd-Navi and some others

6.5.1.3 Spatial Offset

We measure the spatial offset between the checkpoint and the follower's current location when he actually reaches the checkpoint. The spatial offset reflects the metric error where a navigation event like turning should happen and is suitable for measuring navigation performance. It is measured directly in Rviz in ROS platform.

We compare the spatial offset of our system with [9, 11]. Note that in our system, after each relocalization, the follower's location is forcibly aligned to the leader's trajectory owing to the optimization. If a relocalization happens at a checkpoint, the spatial offset will be set as zeo for calibration. If our system receives more sequences, there might be more opportunities of relocalization and more spatial offsets will become zero. So we only conduct experiments on a freshly built map without much trajectory overlapping or the follower's enrichment to the map. The results are shown in Fig. 6.7.

In normal environments with relatively rich texture, i.e., the shopping mall, the supermarket, and the teaching building, our system outperforms others. In all cases, our system has a spatial offset of less than 3.2 m, and in 90% of the cases, the spatial offset is less than 1.1 m. Comparing to magnetic navigation methods *FollowMe* and *FollowUs*, our system shows better accuracy. The spatial offset of *Pair-Navi* is not shown here because the whole idea of *Pair-Navi* is based on relocalization, and therefore it has zero spatial offset.

6.5.1.4 Extreme Conditions

We further experiment the navigation accuracy in an extreme condition, namely inverse directions. Again, we use spatial offset to measure system performance. The result is shown in Fig. 6.8.

We build several long sequences with checkpoints, from 30 m to 150 m, and ask followers to walk in the inverse direction with the sequences and measure the spatial offset at the checkpoints. Since there exist some textureless scenes, the overall performance is plagued by large errors. But for 80% cases, the single way performance is comparative to *FollowMe* and *FollowUs*. Considering in a mature global map, 30 m is a very large length for any passages, so the overall performance in inverse directions is acceptable.

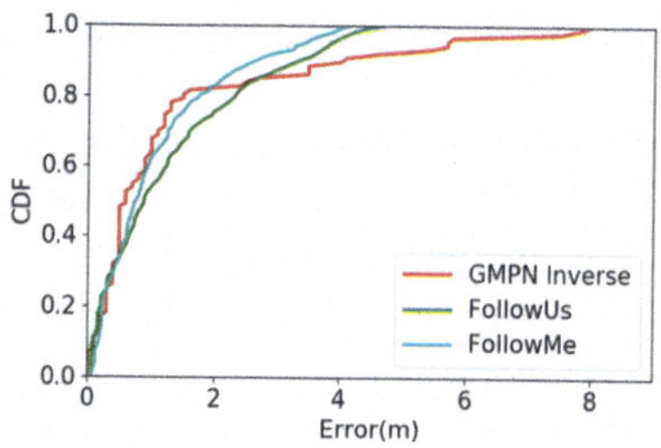

Fig. 6.8 Spatial offset on inverse passages

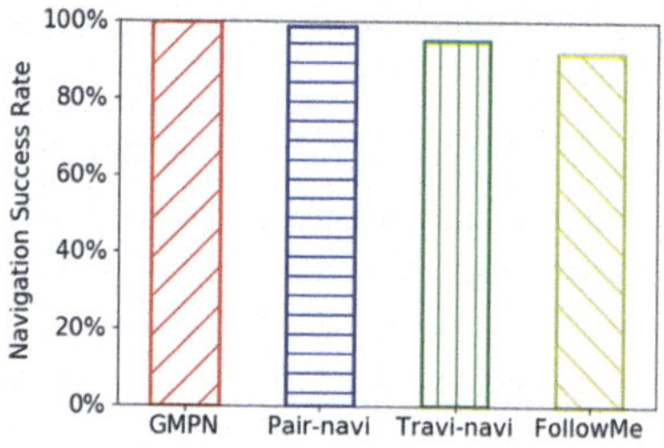

Fig. 6.9 Navigation success rate of our system and several other systems

Comparing to *Pair-Navi* that is not able to work in inverse directions, our system achieves far better applicability. Not to mention that with our crowdsourcing strategy, the performance will become better and better, for when more and more paths are contributed to the global map, our system will be able to detect more and more relocalizations rather than accumulate drift.

6.5.1.5 Navigation Success Rate

We test navigation success rate, which is defined as the rate of successful arrival at the checkpoints. We set 143 checkpoints in 40 navigation paths and let followers go to their destinations without apprising of them the location of the destination or the checkpoints. The results are shown in Fig. 6.9. With the help of our system, all the followers arrive at their destinations and successfully pass 100% of the checkpoints, which is better than *Pair-navi* (98.6%), *Travi-navi* (96%), and *FollowMe* (92%). Even in textureless environments where the spatial offsets were large, the followers managed to pass the checkpoints because the textureless environments are often long aisles or staircases that do not have option paths. Here, we do need to point out some special cases in the shopping mall where the upward and downward escalators between two floors are placed in opposite sides such that the follower cannot follow the same path when he travels in the inverse direction. However, he passes the escalator and continues on the right track anyway and finally makes it to the destination, because from the visualization of the global map, he understands that he should take the escalator on the opposite side. Therefore, these cases are counted as successful pass of the escalator checkpoints.

6.5.1.6 Battery Consumption

We also test battery consumption on smartphones. A Samsung Galaxy S10+ ran for 2 h 35 min, and the energy decreased from 100% to 65%. A Huawei P10 ran for 2 h 9 min, and the energy decreased from 62% to 32%. Considering that a typical indoor navigation takes about 10 min, the battery consumption is acceptable.

6.5.2 *Mobile-Edge Communication Performance*

We study the influence of feature transmission rate on three aspects: the VIO accuracy, the image preprocessing time, and the bandwidth usage. The feature transmission rate means the proportion of images whose feature points are transmitted; in other words, given a transmission rate $0 < r < 1$, the number of all frames n, the number of frames that have their feature points transmitted is $n_0 = n \cdot r$. To avoid uncertainty induced by network congestion, sudden light change, clutter of pedestrian motion, etc. during manual experiment, we conduct this part of experiments by simulation on EuRoc dataset [23]. We choose MH01 (easy), MH03 (medium), and MH05 (difficult) in EuRoc Dataset for varying motions and lighting conditions. The simulation runs on a laptop of Intel Core i7-8750H CPU and 8 GB RAM, with Ubuntu 18.04.

6.5.2.1 On VIO Accuracy

First, as for the influence of transmission rate on VIO accuracy, we use EVO [24] to perform Umeyama alignment on trajectories and measure the overall accuracy. Figure 6.10 shows the RMSE (root mean square error) of tracked trajectories with varying transmission rates on the three datasets. Though VIO accuracy varies as environment and pedestrian motion change, all trajectories' RSMEs are no more than 0.2 m, which is sufficient for indoor navigation as paths usually measure several meters wide. Note also that in challenging scenarios (MH03 and MH05), there is no longer significant improvement on tracking accuracy when transmission rate exceeds 0.2. This indicates that in our mobile-edge architecture, we can limit the transmission overhead to be very minor while keeping the accuracy of VIO satisfactory, i.e., only transmitting feature points for 20% of the frames.

6.5.2.2 On Image Processing Time

Second, as for the influence of transmission rate on image processing time, the results are shown in Fig. 6.11. As aforementioned, VIO accuracy may degrade when more frames are skipped in transmission, but the time cost for image preprocessing may also decrease. This is because if we transmit feature points for a frame, we do

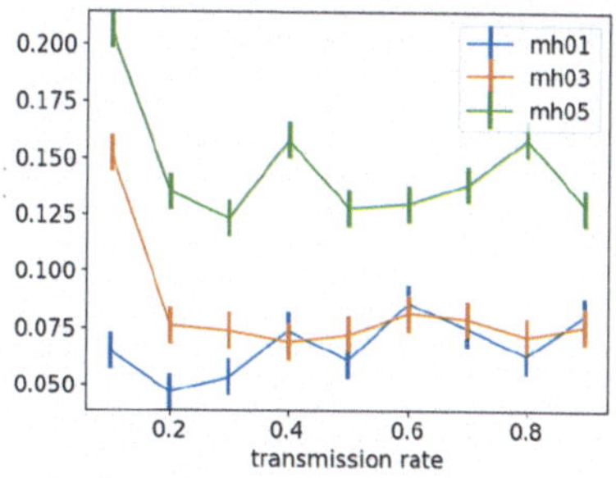

Fig. 6.10 RMSE of trajectories

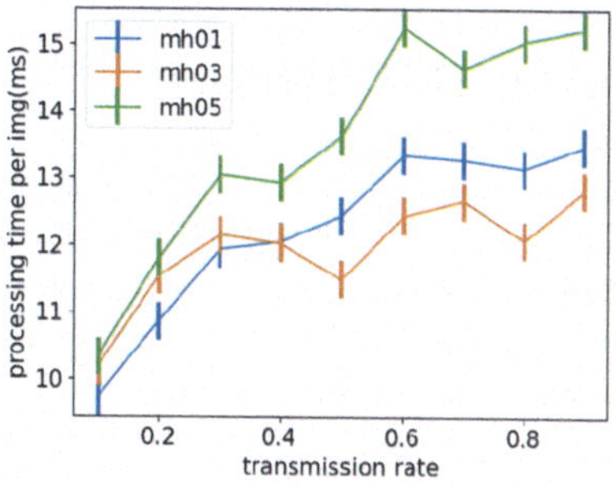

Fig. 6.11 Image processing time

not need to perform fundamental culling, and complementary extracting for it (refer to Sect. 6.4.1, paragraph "Pushing the Limit"). We can see that average image pre-processing including tracking, fundamental culling and complementary extracting, increases as transmission rate changes, though not dramatically. Nonetheless, in the most challenging case (MH05), the average time cost increased by about 1.5 times due to more motion clutter and light changing.

6.5.2.3 On Bandwidth

As for bandwidth, since the major consumption of bandwidth is caused by raw image transmission (raw images are for relocalization, refer to Sect. 6.4.1, paragraph "The mobile side"), we estimate bandwidth usage by dividing the total memory size of raw image transmitted by overall transmission time. In this experiment, we transmit raw images together with feature points. And the frequency of raw image transmission is limited by the frequency of feature transmission. Figure 6.12 demonstrates an almost linearly increasing bandwidth usage which reaches 20 MB/s when the transmission rate is 0.9. Note that although we used a simplified experiment, this experiment result should be similar to practical application; in other words, the bandwidth grows linearly with transmission rate in practice. Therefore, according to this linear growth of bandwidth, we set the transmission of raw images carefully, apart from feature point transmission, and use lossless image compression to rigorously control bandwidth usage.

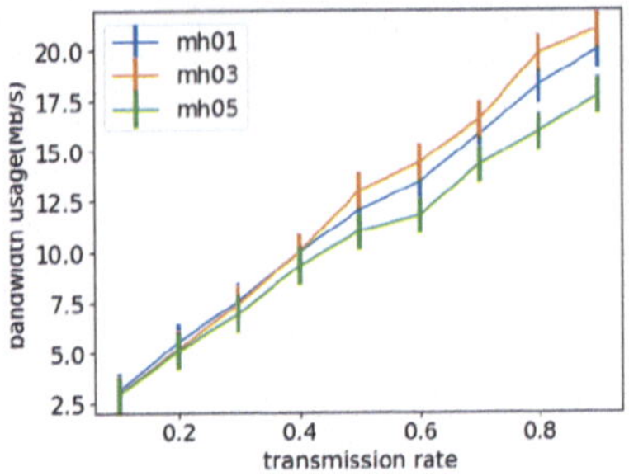

Fig. 6.12 Bandwidth usage

6.5.2.4 Field Experiment

In our field experiment on navigation performance, we observe that the IMU buffer size (shown in the top left of Fig. 6.1) on mobile client averages 30. Note that the Samsung Galaxy S10+ we use generates IMU data faster than 300 Hz. We also find empirically that the processing delay at the edge side is neglectable due to richer computation resources. Ignoring data buffering at routers, Crowd-Navi is able to perform localization and mapping with end-to-end delay at $\frac{30}{300} = 100\,\text{ms}$, indicating it can navigate users in real-time.

6.6 Related Work

6.6.1 Indoor P2P Navigation

Traditional indoor navigation systems navigate users in a global map with a floor plan and infrastructures (e.g., Wi-Fi [3], Bluetooth, RFID [4], etc.). On the contrary, P2P navigation, such as *Travi-Navi* [8], *FollowMe* [9], and *ppNav* [7], does not rely on a global map but rather navigates users in relevant local maps contributed by followers. *Pair-Navi* [10] exploits the power of vision, which is infrastructure-free and efficient. *FollowUs* [11] first exploits the idea of splicing trajectories and obtained more trajectories. However, its localization method may be less than ideal because a follower first needs to walk for some distance to acquire a segment of magnetometer readings for segment matching, but visual methods can perform immediate localization by visual relocalization. In addition, the comprehensiveness of path coverage, though seemingly fundamental at first glance, is hardly discussed by previous works like [11] that adopt such a scheme. In our work, we exploit the idea of merging local maps with visual approaches and give a theoretical proof and experimental evaluation of the large path archive.

6.6.2 *Visual-Inertial Odometry*

Works on visual-inertial odometry can be classified into two categories, the loosely coupled [25] and the tightly coupled [26]. The loosely coupled is comparatively lightweight and efficient, but the accuracy may be unsatisfactory in our applications where a global consistent map relies on the accuracy of each local map. The tightly coupled has several prototypes, including EKF-based approaches [27] and optimization-based approaches [14]. With the advance of preintegration methods [14], IMU observations can be optimized together with visual observations, enabling complete systems to be more accurate, among which *VINS-Mono* [21] is one of the state-of-the-art systems. Though *VINS-Mono* [21] achieves satisfactory accuracy, it cannot run in most smartphones in real-time without SACI, especially Android phones. We modify the system and extend it into a mobile-edge structure to assure real-time performance without sacrificing accuracy.

6.7 Conclusion

We present a visual peer-to-peer navigation system that navigates in a global map constructed through a profound crowdsourcing scheme that merges local maps together. To the best of our knowledge, we are the first to give a theoretical derivation of the path abundance of this crowdsourcing scheme as well as an experimental validation. The system is powered by a mobile-edge architecture with visual-inertial odometry. It overcomes the scale ambiguity and the unidirectional defect of visual paths and thus achieves better applicability than existing peer-to-peer navigation works. Experiments show that the navigation performance of our navigation system is also delightful. Future works may include improving the VIO system for better performance in textureless environments.

Appendix: Path Coverage Model

We present a quantified argument on the impact of splicing independent sequences upon increasing total path coverage. We use a chessboard model as an abstraction of a typical indoor floor plan and give a lower bound of the possibility of two independent sequences intersecting. Then we use this lower bound of possibility to derive the expectation of the number of paths connecting two places. We find this expectation grows faster than any polynomial with regard to the amount of available sequences (m) in the sparse case where m is far less than the amount of POIs.

We define a chessboard to be a set of 2D points uniformly distributed on a plane (vertices) $V = \{(u, v)|u = x_0 + t, v = y_0 + s, t \leq n, s \leq m, t, s \in \mathbb{N}\}$ with edges connecting only nearest neighbors $E = \{\{(u_1, v_1), (u_2, v_2)\}|\max\{|u_1 -$

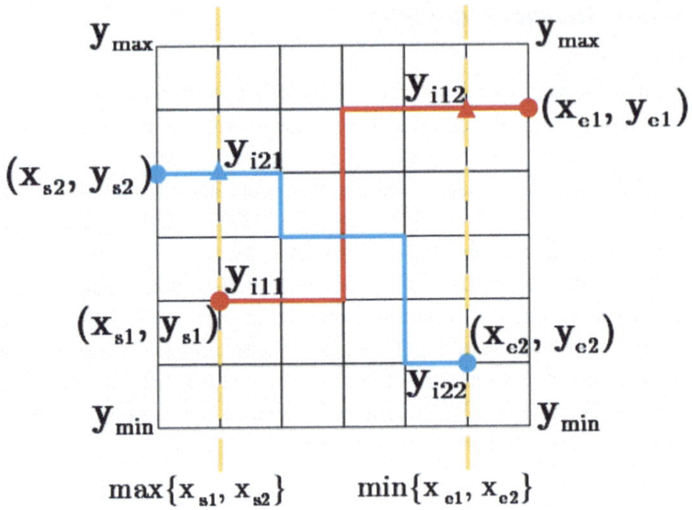

Fig. 6.13 Intersection of two paths on a chessboard

$u_2|, |v_1 - v_2|\} = 1, \min\{|u_1 - u_2|, |v_1 - v_2|\} = 0\}$. Its visualization is in Fig. 6.13. Most floor plans consist of only two perpendicular directions of passages and thus can be embedded into a chessboard by regarding crossings and POIs as vertices and hallway as edges. Given two vertices $A(x_s, y_s), B(x_e, y_e)$ $(x_s \leq x_e)$, a *path* $p(x_s, y_s, x_e, y_e)$ between A and B is one of the shortest series of adjacent edges starting at A and ending at B. We consider only the shortest ones, so that the paths are finite, otherwise there may be infinite loops. We assume $(x_s, y_s), (x_e, y_e)$ independently conform to the uniform distribution among the chessboard vertices, and all paths between them are of equal possibility when x_s, y_s, x_e, y_e are fixed. Thus a distribution of a path can be derived.

Given two independently selected paths $p_1(x_{s1}, y_{s1}, x_{e1}, y_{e1})$ and $p_2(x_{s2}, y_{s2}, x_{e2}, y_{e2})$, a necessary condition for them to intersect is $\max\{x_{s1}, x_{s2}\} \leq \min\{x_{e1}, x_{e2}\}$, which means their projections on x-axis intersect. Let x_{min} and x_{max} be the minimum and maximum x components of all vertices. We first consider a large chessboard (m and n in vertices' definition are big) and define $\phi(x) = \frac{x - x_{min}}{x_{max} - x_{min}}$, and we may assume that $\phi(x_{s1}), \phi(x_{e1}), \phi(x_{s1}), \phi(x_{e2}) \sim U[0, 1] i.d.d.$ The complement of all possible choices of $(\phi(x_{s1}), \phi(x_{e1}), \phi(x_{s2}), \phi(x_{e2}))$ to make $\max\{x_{s1}, x_{s2}\} \leq \min\{x_{e1}, x_{e2}\}$ forms a union of two pyramids $E_1 \cup E_2$ within a cube $I^4, I = [0, 1]$, where $E_1 = \{(x_1, x_2, x_3, x_4) | \max\{x_1, x_2\} \leq \min\{x_3, x_4\}\}$ and $E_2 = \{(x_1, x_2, x_3, x_4) | \min\{x_1, x_2\} \geq \max\{x_3, x_4\}\}$, with $x_i \in [0, 1], i = 1, 2, 3, 4$. By calculating the pyramids' volume, we get the possibility for two paths' projections intersecting on x-axis to be $\frac{7}{9}$.

Under the condition that two paths' projections on x-axis intersect, for each of the paths, let $\mathbf{y_i} = [y_{i1}, y_{i2}]^T, i = 1, 2$, be a vector whose components are the path's intersection with lines $x = \max\{x_{s1}, x_{s2}\}$ and $x = \min\{x_{e1}, x_{e2}\}$. Let $y_{\max}$ and $y_{\min}$

be the maximum and minimum y components of all vertices. We define the *relative y-pair*, to be the vector $\mathbf{y_r} = [y_{r1}, y_{r2}]$ satisfying $y_{rj} = \frac{y_{ij} - y_{\min}}{y_{\max} - y_{\min}}$, $j = 1, 2$. Figure 6.13 gives an illustration of all the notations defined above. When the chessboard is large, we consider y_{r1} and y_{r2} to be real variables on [0, 1] whose joint probability density function is ρ. Let $\mathbf{y_{r1}} = [u_1, v_1]^T$ and $\mathbf{y_{r2}} = [u_2, v_2]^T$ be the relative y-pairs of p_1 and p_2; then $\mathbf{y_{r1}}$ and $\mathbf{y_{r2}}$ are independent and identically distributed since p_1 and p_2 are selected independently. The chessboard's symmetry on x and y directions assures that

$$\rho(u, v) = \rho(v, u) = \rho(1 - u, 1 - v) = \rho(1 - v, 1 - u). \tag{6.5}$$

According to the Existence Theorem of Zero Point of continuous functions on a closed interval, p_1 and p_2 intersect when $(u_1 - u_2)(v_1 - v_2) < 0$, and we consider its probability P to be our desired lower bound. Given $(u, v) \in I^2$, where $I = [0, 1]$, let $E_{(u,v)} = \{(u', v') | (u - u')(v - v') < 0, (u', v') \in I^2\}$, which is the two shadowed areas in Fig. 6.14.We then have

$$P = \iint_{I^2} \left(\iint_{E_{(u,v)}} \rho(s, t) ds\, dt \right) \rho(u, v) du\, dv. \tag{6.6}$$

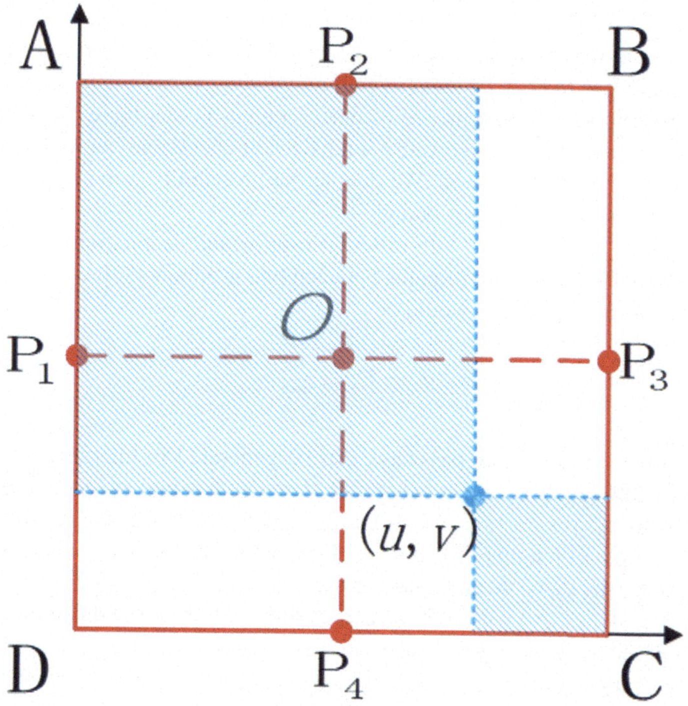

Fig. 6.14 Integration area of ρ and an example of $E_{(u,v)}$

Because the four squares are mapped bijectively by the symmetry in Eq. (6.5), we then have

$$\begin{aligned}\iint_{OP_1DP_4}\rho(u,v)du\,dv &= \iint_{OP_1AP_2}\rho(u,v)du\,dv\\ &= \iint_{OP_2BP_3}\rho(u,v)du\,dv = \iint_{OP_3CP_4}\rho(u,v)du\,dv.\end{aligned} \tag{6.7}$$

Therefore, each of the above integral would be $\frac{1}{4}$. Note that $(u,v)\in OP_1AP_2 \Rightarrow E_{(u,v)} \supset OP_3CP_4$ and $(u,v)\in OP_3AP_4 \Rightarrow E_{(u,v)} \supset OP_1CP_2$. So we have

$$\begin{aligned}&\iint_{I^2}\left(\iint_{E_{(u,v)}}\rho(s,t)ds\,dt\right)\rho(u,v)du\,dv\\ &> \iint_{OP_1AP_2}\left(\iint_{OP_3CP_4}\rho(s,t)ds\,dt\right)\rho(u,v)du\,dv\\ &+ \iint_{OP_3CP_4}\left(\iint_{OP_1AP_2}\rho(s,t)ds\,dt\right)\rho(u,v)du\,dv\\ &= \frac{1}{4}\cdot\frac{1}{4}+\frac{1}{4}\cdot\frac{1}{4}=\frac{1}{8}.\end{aligned} \tag{6.8}$$

Therefore we derive the lower bound of the probability of two paths' intersecting in the asymptotic case to be $\frac{1}{8}\cdot\frac{7}{9}=\frac{7}{72}$.

The most important consequence of our argument is that there indeed exists a constant lower bound for intersecting for all sizes of chessboard as we have proved the case when m and n go to ∞. We denote the probability to be P_0 and keep in mind that it has a constant lower bound.

Given n POIs and m paths with different origins and destinations, we consider the expectation of the amount of paths between two arbitrary POIs a and b. In pure P2P systems, this expectation would be

$$E = E_1 = \frac{m}{C_n^2}. \tag{6.9}$$

However, after splicing, new passways can be generated by merging a sequence of original paths $(p_1, p_2, ..., p_i)$, where p_1 starts from a and p_n ends at b and p_k intersects with $p_{k+1}, k = 1, ..., i-1$. Let E_i denote the expectation of the amount of such sequences of length i. To compute E_i, first fix i pair of POIs: $(v_{k1}, v_{k2}), k = 1, 2, ..., i, v_{11} = a, v_{22} = b$, and consider the probability of existence of a sequence of i paths $(p_1, p_2, ..., p_i)$ with each p_k connecting (v_{k1}, v_{k2}). Let this probability be P_i, and let N_i be the amount of all such sequence of paths as $v_{kj}, k = 1, ..., i$ (except v_{11} and v_{i2}) vary within $(n-2)$ POIs other than a and b. Then $E_i = P_iN_i$. P_i is derived by multiplying the probability of the event where all of the pairs (v_{k1}, v_{k2})

mentioned above are indeed connected and that p_k intersects with p_{k+1}. Thus,

$$P_i > \frac{\prod_{k=0}^{i-1}(m-k)}{\prod_{k=0}^{i-1}(C_n^2-k)} P_0^{i-1}; N_i = (n-2)^2 A_{C_{n-2}^2}^{i-2}. \tag{6.10}$$

Therefore,

$$\begin{aligned} E = \sum_{i=1}^{m} E_i &> \frac{m}{C_n^2} + \sum_{i=2}^{m} \left(P_0^{i-1}(n-2)^2 A_{C_{n-2}^2}^{i-2} \frac{\prod_{k=0}^{i-1}(m-k)}{\prod_{k=0}^{i-1}(C_n^2-k)} \right) \\ &\approx \frac{m}{C_n^2} + \sum_{i=2}^{m} \left(\frac{4}{n^2} P_0^{i-1} \prod_{k=0}^{i-1}(m-k) \right), \end{aligned} \tag{6.11}$$

where E_i is the expectation of the amount of available pathways of length i. The approximation is made by assuming n to be very big. Note that all terms has a constant coefficient divide by n^2. The last term grows with the speed of $O(m!)$, faster than any polynomial. As a result, one could expect the amount of pathways connecting two arbitrary POIs quickly exceeds 1 as m increases from bootstrap, providing better path coverage than pure P2P navigation systems.

References

1. Z. Yang, C. Wu, Y. Liu, Locating in fingerprint space: wireless indoor localization with little human intervention, in *Proceedings of the ACM MobiCom* (2012)
2. C. Wu, Z. Yang, Y. Liu, Smartphones based crowdsourcing for indoor localization. IEEE Trans. Mob. Comput. **14**(2), 444–457 (2015)
3. H. Liu, Y. Gan, J. Yang, et al., Push the limit of WiFi based localization for smartphones, in *Proceedings of the ACM MobiCom* (2012)
4. J. Wang, D. Katabi, Dude, where's my card? RFID positioning that works with multipath and non-line of sight, in *Proceedings of the ACM SIGCOMM* (2013)
5. L. Shangguan, Z. Yang, A.X. Liu, Z. Zhou, Y. Liu, STPP: spatial-temporal phase profiling-based method for relative RFID tag localization. IEEE/ACM Trans. Networking **25**(1), 596–609 (2017)
6. M. Azizyan, I. Constandache, R. Roy Choudhury, SurroundSense: mobile phone localization via ambience fingerprinting, in *Proceedings of the ACM MobiCom* (2009)
7. Z. Yin, C. Wu, Z. Yang, Y. Liu, Peer-to-peer indoor navigation using smartphones. IEEE J. Sel. Areas Commun. **35**(5), 1141–1153 (2017)
8. Y. Zheng, G. Shen, L. Li, C. Zhao, M. Li, F. Zhao, Travi-navi: self-deployable indoor navigation system, in *Proceedings of the ACM MobiCom* (2014)
9. Y. Shu, K.G. Shin, T. He, J. Chen, Last-mile navigation using smartphones, in *Proceedings of ACM MobiCom* (2015)
10. E. Dong, J. Xu, C. Wu, Z. Yang, Y. Liu, Pair-navi: peer to peer indoor navigation with mobile visual slam, in *Proceedings of the IEEE INFOCOM* (2019)
11. Y. Shu, Z. Li, B. Karlsson, Y. Lin, T. Moscibroda, K. Shin, Incrementally-deployable indoor navigation with automatic trace generation, in *IEEE INFOCOM 2019-IEEE Conference on Computer Communications* (IEEE, New York, 2019), pp. 2395–2403

12. L. Liu, H. Li, M. Gruteser, Edge assisted real-time object detection for mobile augmented reality, in *Proceedings of the ACM MobiCom* (2019)
13. T.Y.-H. Chen, L. Ravindranath, S. Deng, P. Bahl, H. Balakrishnan, Glimpse: continuous, real-time object recognition on mobile devices, in *Proceedings of the ACM Sensys* (2015)
14. C. Forster, L. Carlone, F. Dellaert, D. Scaramuzza, On-manifold preintegration for real-time visual–inertial odometry. IEEE Trans. Robot. **33**(1), 1–21 (2016)
15. R. Mur-Artal, J.D. Tardós, Visual-inertial monocular slam with map reuse. IEEE Rob. Autom. Lett. **2**(2), 796–803 (2017)
16. T. Qin, S. Shen, Robust initialization of monocular visual-inertial estimation on aerial robots, in *2017 IEEE/RSJ International Conference on Intelligent Robots and Systems (IROS)* (IEEE, New York, 2017), pp. 4225–4232
17. R. Hartley, A. Zisserman, *Multiple View Geometry in Computer Vision* (Cambridge University Press, Cambridge, 2003)
18. V. Lepetit, F. Moreno-Noguer, P. Fua, EPnP: An accurate O(n) solution to the PnP problem. Int. J. Comput. Vis. **81**(2), 155 (2009)
19. T. Qin, P. Li, S. Shen, Relocalization, global optimization and map merging for monocular visual-inertial slam, in *2018 IEEE International Conference on Robotics and Automation (ICRA)* (IEEE, New York, 2018), pp. 1197–1204
20. B.A. Galler, M.J. Fisher, An improved equivalence algorithm. Commun. ACM **7**(5), 301–303 (1964)
21. T. Qin, P. Li, S. Shen, VINS-mono: a robust and versatile monocular visual-inertial state estimator. IEEE Trans. Robot. **34**(4), 1004–1020 (2018)
22. *ROS kinetic kame*. http://wiki.ros.org/kinetic
23. M. Burri, J. Nikolic, P. Gohl, et al., The EuRoC micro aerial vehicle datasets. Int. J. Robot. Res. **35**(10), 1157–1163 (2016)
24. E. script. https://github.com/MichaelGrupp/evo.
25. S. Lynen, M.W. Achtelik, S. Weiss, M. Chli, R. Siegwart, A robust and modular multi-sensor fusion approach applied to MAV navigation, in *2013 IEEE/RSJ International Conference on Intelligent Robots and Systems* (IEEE, New York, 2013), pp. 3923–3929
26. S. Leutenegger, P. Furgale, V. Rabaud, M. Chli, K. Konolige, R. Siegwart, Keyframe-based visual-inertial slam using nonlinear optimization, in *Proceedings of Robotics Science and Systems (RSS) 2013* (2013)
27. A.I. Mourikis, S.I. Roumeliotis, A multi-state constraint Kalman filter for vision-aided inertial navigation, in *Proceedings 2007 IEEE International Conference on Robotics and Automation* (IEEE, New York, 2007), pp. 3565–3572

Chapter 7
Environment Understanding with EdgeSLAM

Abstract Accurate and efficient instance segmentation on mobile platforms is crucial for applications in augmented reality, inspection tasks, and understanding environments. Despite the development of edge-assisted systems to facilitate computation-intensive operations, the unique challenges of instance segmentation, including device and target motion, precision requirements, and computational burdens, limit its performance on resource-constrained devices.

This chapter introduces edgeIS, an innovative system combining mobile and edge computing resources to deliver real-time, precise instance segmentation. edgeIS leverages mobile device capabilities to sense environmental dynamics and its motion, fostering a novel collaboration between mobile devices and edge servers specifically tailored for segmentation tasks. Implemented across various mobile devices and a lightweight edge server, edgeIS undergoes rigorous testing across three datasets, demonstrating its capability to operate in real-time with a segmentation IoU of 0.92, surpassing current leading solutions. Moreover, edgeIS's integration into an augmented reality inspection system within an oil-field showcases its industrial applicability and performance excellence.

7.1 Introduction

Instance segmentation, pivotal for myriad applications like robotics and content retrieval, involves pinpointing and delineating each distinct object within an image, as detailed in [1]. The advent of mobile technologies such as smartphones and drones has spurred the necessity for executing real-time instance segmentation, a capability integral to augmented reality (AR, as shown in Fig. 7.1) [2], autonomous navigation [3], and context-aware services [4]. In scenarios where immediate object and human recognition is paramount—for instance, robots operating in intricate environments or for rendering immersive AR experiences [5]–the demand for swift and accurate segmentation is especially high. Nevertheless, the complexity of deep learning (DL) models for such tasks, typically requiring high-end GPUs for a blend of precision and speed [6], poses a substantial challenge for deployment on conventional mobile devices. Despite strides in DL model optimization [7, 8],

J. Xu et al., *Edge Assisted Mobile Visual SLAM*,
https://doi.org/10.1007/978-981-97-3573-0_7

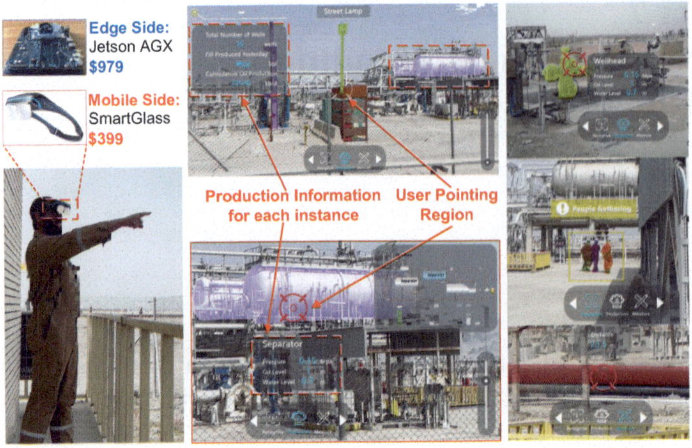

Fig. 7.1 The user interface of edgeIS is showcased within an AR framework for industrial checks at one of the globe's expansive oil-fields. edgeIS stands out by delivering **immediate** and **precise** instance segmentation on portable devices, powered by an edge node designed for limited resources. As users focus on specific regions within their viewpoint, edgeIS identifies and annotates those areas (e.g., separators, pipes) with relevant industrial data. This setup underscores the importance of edgeIS's exactness and swift response: Inaccuracies in identifying a mere few pixels can result in crucial data being incorrectly positioned, while the dynamic nature of the environment and user movement necessitates minimal delay in processing

achieving real-time instance segmentation inference on these devices remains elusive.

In recent advancements, the integration of edge computing into research has notably improved the feasibility of executing real-time applications that demand substantial computational power. By leveraging the capabilities of edge servers or nodes, tasks that require intensive computation but are less sensitive to delays can be offloaded from mobile devices to the edge. This approach allows mobile devices to concentrate on executing tasks that, although less computationally demanding, require immediate processing [9]. Among the leading-edge methodologies, Edge-Duet [10] and EAAR [11] have utilized the edge computing paradigm to enhance the accuracy of object detection tasks on mobile devices without compromising real-time performance. These methods primarily employ a sophisticated "track+detect" mechanism [12], where the device tracks objects in a new frame using previously cached detection results and sends only essential frames to the edge for a detailed analysis and updates, as shown in Fig. 7.2a.

Motivated by contemporary advancements, a pertinent question arises: Is it feasible to utilize an edge-supported framework to facilitate real-time segmentation on handheld devices? In pursuit of an answer, we collaborated with a leading enterprise to incorporate this edge-based system into their oil-field inspection industrial processes. However, the adoption of the traditional "track+detect" model did not yield satisfactory results for segmentation tasks. The limitations encountered

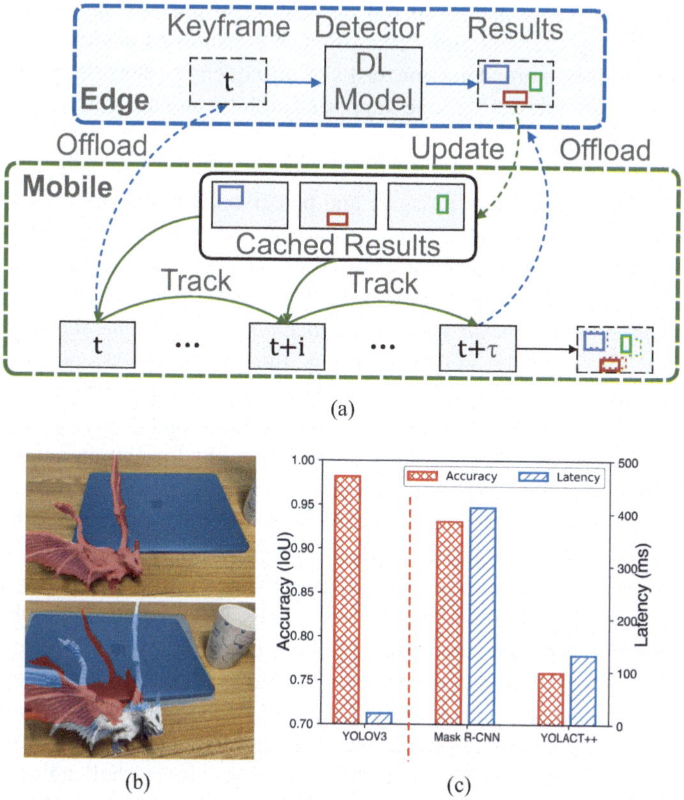

Fig. 7.2 Overview of the "track+detect" framework and challenges in simply applying it to instance segmentation tasks. (**a**) The architecture of the framework. (**b**) The object tracker designed for detection tasks cannot accurately predict the segmentation masks on mobile devices, especially for those moving objects (e.g., dragon in the figure). (**c**) Unlike detection tasks, it is still challenging for a commercial edge node to achieve both high-precision and real-time performance on segmentation tasks. (**a**) Illustration of the "track+detect" framework. (**b**) On-device predictions. (**c**) Performance of models on the edge

are primarily twofold: Firstly, segmentation tasks demand an accurate pixel-wise delineation of objects, a more refined requirement than the approximate bounding boxes needed for detection. The existing object tracking mechanisms [10, 11, 13] fall short in transmitting detailed masks across consecutive frames with the necessary precision, as illustrated in Fig. 7.2b and further analyzed in Sect. 7.6.3. Feedback from the field personnel highlighted that even minor discrepancies of a few pixels could misplace the mask onto an incorrect object, leading to erroneous semantic classification of the targeted region. Secondly, attaining real-time and accurate instance segmentation on edge servers is hindered by significant computational demands. Moreover, in actual industrial settings, there is a preference for compact and less power-consuming edge devices (e.g., Nvidia Jetson TX2

[14], AGX Xavier [15]) due to spatial and electrical constraints. As depicted in Fig. 7.2c, executing segmentation tasks on such constrained devices proves to be more daunting than mere detection tasks. Consequently, there is an imperative need for devising an innovative edge-supported architecture tailored for the instance segmentation challenge.

In the current technological landscape, mobile devices are equipped with the functionality to autonomously navigate and perceive their surroundings. Platforms like ARCore and ARKit, for instance, harness the power of visual odometry (VO) [16] and visual-inertial odometry (VIO) [17] to ascertain a smartphone's position and orientation. This discovery leads us to the potential of *incorporating VO within an edge-supported framework to facilitate real-time segmentation on smartphones*. Theoretically, VO's output (e.g., the mobile device's positional shifts and the movement of objects within view) could serve as a foundation to *propagate* segmentation masks across successive frames. By accurately overlaying the mask on the object, the edge's deep learning algorithm could concentrate on a narrower vicinity of the object, thus expediting the mask *deduction* procedure.

Nonetheless, materializing this concept into a viable system presents several significant hurdles. Initially, the transition of pixel-level masks on mobile platforms poses a challenge. To accurately shift instance masks from one frame to the next, it is crucial to track pixel movements, particularly those outlining the object. Yet, while current VO techniques construct a 3D representation through feature points, these points are notably less dense than the image's pixels, and it is indeterminate whether these points are part of the object [18]. Hence, tracking is limited to a minimal fraction of the object's pixels, resulting in imprecise mask forecasts.

Moreover, leveraging predictive results on the edge proves challenging. Within a deep learning framework, various layers exchange information via feature maps. Consequently, the straightforward, human-intelligible object location predictions generated by mobile platforms are insufficient for guiding the model to omit nonessential computations at the feature map level.

Finally, minimizing the volume of data sent poses a considerable challenge. In architectures assisted by edge computing, the delay caused by data transmission critically impacts the system's efficiency [19]. While existing studies have extensively explored prioritizing and transmitting different image regions based on prior outcomes, these methodologies falter in scenarios with dynamic object and device positions. Without precise motion data, previous approaches struggle to pinpoint detailed regions of interest, such as potential object locations or newly visible scenes, thereby failing to effectively minimize data transmission.

To address the aforementioned obstacles, we have developed and implemented edgeIS, marking the inception of an **edge**-oriented architecture that facilitates real-time **I**nstance **S**egmentation on handheld devices. edgeIS revolutionizes the classic "track+detect" model typical of edge-enhanced frameworks, introducing an innovative "transfer+infer" strategy for cooperation between mobile units and the edge. This strategy capitalizes on the visual detection capabilities of mobile devices to *transfer* previously segmented masks, concurrently utilizing these predictive outcomes to expedite *inference* processes at the edge.

Within edgeIS, initially, a *motion aware mobile mask transition* component is employed to seamlessly map the segmentation outcomes from an antecedent keyframe onto a subsequent frame, leveraging motion insights of both the device and the object in question. Furthermore, a *contour-driven edge inference optimization* approach is crafted, minimizing redundant calculations and enhancing the inference speed on the edge server, guided by mobile-derived forecasts. Notably diverging from the conventional unidirectional support of the edge for mobile devices, this innovative approach adeptly harnesses mobile-originated insights to amplify the efficiency of the edge's deep learning algorithms. In addition, we introduce a *content-determined fine-grained RoI delineation* method aimed at optimizing the volume of data transmitted. By leveraging prior knowledge alongside a spatial comprehension of the prevailing scene, edgeIS achieves a reduction in transmission delays without detriment to operational efficacy on either front.

We fully implement edgeSLAM's server side on an Nvidia Jetson TX2 and mobile side on two different types of mobile devices. We conduct comprehensive experiments under different network conditions and examine edgeIS on two public datasets and a self-labeled dataset with more than 19*k* frames. We also compare edgeIS with two state-of-the-art related works, EdgeDuet [10] and EAAR [11]. The results show that edgeIS achieves an average frame rate of 30 fps with 0.92 segmentation IoU (intersection over union), improving the accuracy by 10–20% and reducing the false rate by at least 70% compared with previous studies.

The key contributions are summarized as follows:

- We introduce edgeIS, to our knowledge, the pioneering solution that empowers mobile devices with the capability to conduct instance segmentation in real-time. This achievement comes from innovatively rethinking and reconstructing the edge-supported architectural design to cater to such a computationally demanding task.
- We successfully incorporate Visual Odometry into our collaborative framework between mobile devices and the edge, establishing a novel "transfer+infer" mechanism. This approach enables a dual-benefit system where not only does the mobile segment transfer outcomes to successive frames, but it also significantly boosts the speed of deep learning model predictions on edge nodes, which are typically limited in computational resources.
- Through comprehensive evaluation and benchmarking against two analogous systems, we demonstrate the exceptional efficacy of edgeIS across various conditions. Additionally, implementing edgeIS within an AR-enhanced inspection application for oil-fields, we validate its segmentation precision and real-time capabilities, confirming its applicability to demanding industrial contexts.

7.2 System Architecture for EdgeSLAM-Based Environment Understanding

Figure 7.3 outlines the architectural blueprint of edgeIS. Viewed from a high level, edgeIS introduces an innovative "transfer+infer" strategy, fostering mutual enhancement between the mobile and edge components. On the mobile front, essential VO functionalities are redefined and seamlessly woven into the *motion aware mobile mask transfer* framework. This allows edgeIS to accurately monitor the device's movement and the dynamics of observed objects, utilizing this information to instantaneously generate segmentation masks for the current frame.

At the edge, the *contour instructed edge inference acceleration* feature plays a crucial role in expediting the computation of the complex DL model. In detail, edgeIS intricately integrates prior mobile-generated predictions into the DL algorithm through strategic *dynamic anchor placement* and *RoI pruning*.

Additionally, a *content-based fine-grained RoI selection* strategy is employed to judiciously determine (i) the necessity of transmitting a frame to the edge and (ii) the focal points within the frame, identifying areas that can be efficiently compressed without significant detail. Thanks to these methodologies, edgeIS markedly minimizes reliance on network bandwidth while maintaining system efficiency and accuracy.

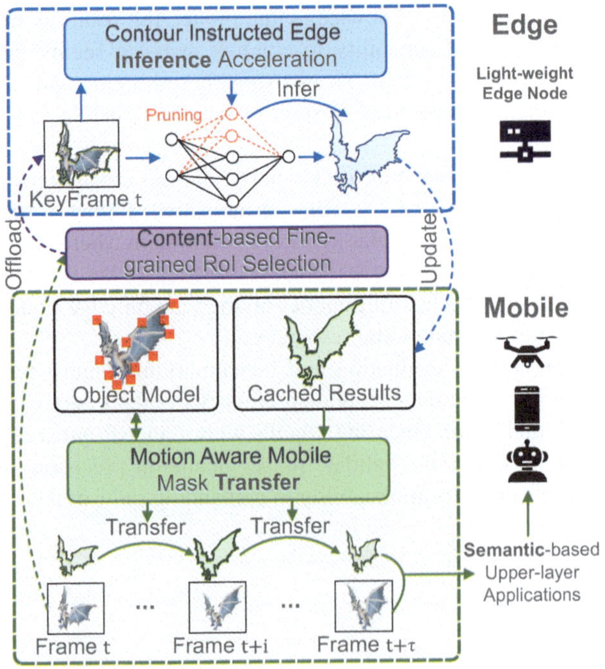

Fig. 7.3 Overall architecture of edgeIS

7.3 Motion Aware Mobile Mask Transfer

In edgeIS, we implement a *motion aware mobile mask transfer* strategy, allowing mobile devices to precisely forecast segmentation outcomes for the present frame in real-time (i.e., aligning with a 30 fps camera rate). Given the challenges posed by the movement of both the target objects and the mobile device, alongside the inherent difficulties of the instance segmentation task, traditional update methods (e.g., employing motion vectors, KCF, or feature matching) often fall short in yielding satisfactory mask prediction results. Instead, by integrating VO and harnessing the understanding of both the objects' and device's motion, we achieve refined pixel-level mask transitions across consecutive frames.

The operational flow of the *mobile mask inference* module is depicted in Fig. 7.4. Initially, edgeIS sets up the mobile device's pose and constructs a 3D model of the observed objects, utilizing sequential frames along with their precise masks received from the edge server. Following this setup, the device and objects' poses are dynamically tracked and updated through a motion tracking system. Subsequently, for every new frame, edgeIS applies spatial insights and preceding masks to deduce the mask for the current pose. Further elaboration on each functional component is provided below.

7.3.1 *Initial Object Modeling*

During initialization, a pair of video frames is selected from continuous input to estimate the starting pose and model the observed objects in 3D space. The two determined frames will be sent to the edge server for masks with accurate semantic annotations generated by deep learning models. To construct the initial 3D map, edgeIS performs feature extraction and matching between the two frames to choose a set of feature pixels representing the same 3D positions in the two frames. In our implementation, we use ORB [20] feature for its efficiency in computing and robustness against the change of viewpoints.

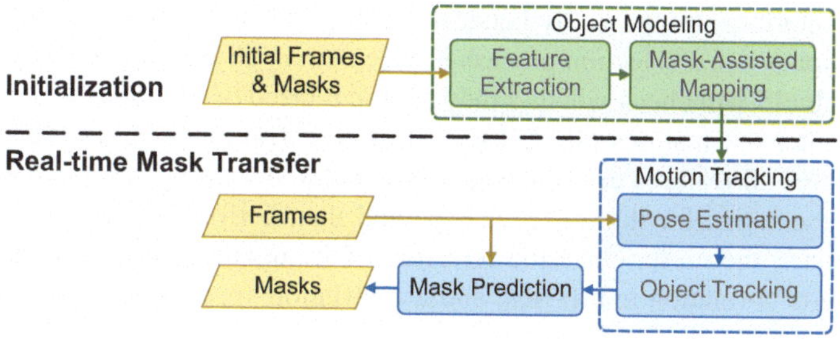

Fig. 7.4 The workflow of the *Motion Aware Mobile Mask Transfer* scheme

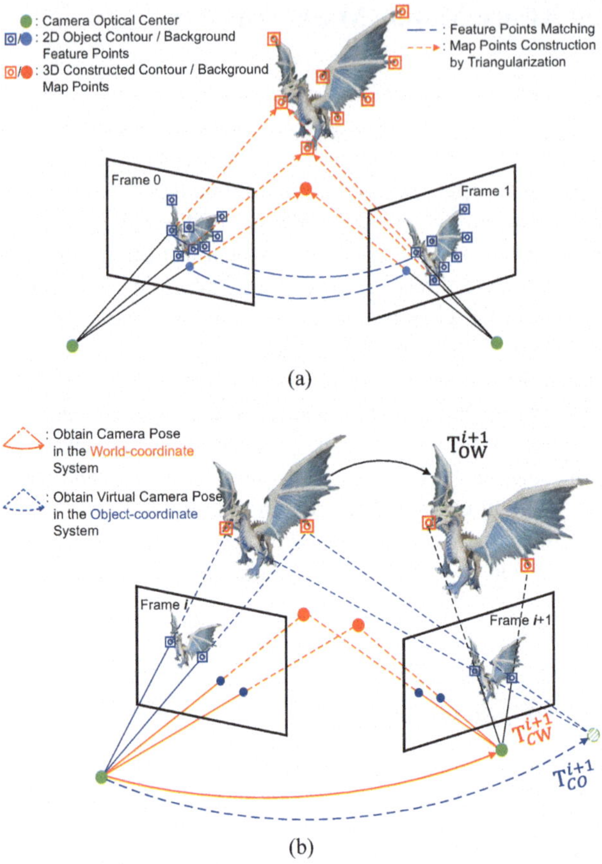

Fig. 7.5 Motion aware mask transfer module. (**a**) Initial object modeling. (**b**) Pose estimation and object tracking

According to accurate masks from the edge server, the features are divided into two groups. As illustrated in Fig. 7.5a, denote the two frames as Frame 0 and Frame 1, any pair of features belonging to the background mask as $\mathbf{p}_0$ and $\mathbf{p}_1$, and other matched features as $\mathbf{q}_0$ and $\mathbf{q}_1$. To better model the objects of interest while avoiding heavy computation, a selection process is executed on all features. For background features, edgeIS will check whether they are too blurred or too close to neighboring ones and filter out features that fail the check. For features within masks, edgeIS first preserves all features near the edge of the mask since pixels on the contour are more representative for the object's shape and then performs blurriness check on features inside the mask. Once the selection of features is done, the relative pose (consisting of a rotation matrix $\mathbf{R}_{10}$ and a translation vector $\mathbf{t}_{10}$) between the two frames is computed by first using epipolar geometry to solve fundamental matrix $\mathbf{F}_{10}$ [21].

$$\begin{aligned} \mathbf{p}_1^{\mathrm{T}} \mathbf{F}_{10} \mathbf{p}_0 &= 0, \\ \mathbf{q}_1^{\mathrm{T}} \mathbf{F}_{10} \mathbf{q}_0 &= 0. \end{aligned} \tag{7.1}$$

Solving $\mathbf{F}_{10}$ requires no less than eight pairs of features and edgeIS will first use all pairs of $\mathbf{p}_0$ and $\mathbf{p}_1$ since the pixels of background are more likely to be static. Then the relative pose between the two frames is solved by

$$\mathbf{t}_{10}^{\wedge} \mathbf{R}_{10} = \mathbf{K}^{\mathrm{T}} \mathbf{F}_{10} \mathbf{K}, \tag{7.2}$$

where $\mathbf{K}$ is the intrinsic matrix of the camera and $(\cdot)^{\wedge}$ calculates the skew-symmetric matrix from a vector. With the recovered pose, VO triangulates 3D points from selected features and constructs the initial map of the captured scene (taking $\mathbf{p}_0$ and $\mathbf{p}_0$ as example):

$$s_1 \mathbf{R}_{10} (\mathbf{K}^{-1} \mathbf{p}_0)^{\wedge} (\mathbf{K}^{-1} \mathbf{p}_1) + \mathbf{t}_{10}^{\wedge} (\mathbf{K}^{-1} \mathbf{p}_1) = 0, \tag{7.3}$$

where s_1 is the depth of features in Frame 1 and s_0 can also be computed using the same method. Once a 3D point is created, edgeIS annotates it according to its corresponding features. If the two features belong to masks with the same label (e.g., people or car), the 3D point will be labeled as the same class and otherwise as background. After initialization, edgeIS obtains a starting pose for continuous motion tracking and an annotated map that models its observed objects.

7.3.2 *Motion Tracking*

As the mobile device moves around, VO continuously tracks its pose and updates the 3D map. For the i-th coming frame, VO first detects feature pixels in it and then matches them to 3D points in the map. With this correspondence from 2D pixels to 3D points, VO can solve the device pose of i-th frame in the world coordinate $\mathbf{T}_{\mathrm{CW}}^{i} = [\mathbf{R}_{\mathrm{CW}}^{i} | \mathbf{t}_{\mathrm{CW}}^{i}]$ by minimizing the reprojection error between the map points projected to the frame given the pose of the frame and the features in it. The minimization can be solved as an optimization problem using bundle adjustment:

$$\mathbf{T}_{\mathrm{CW}}^{i} = \arg\min_{\mathbf{T}_{\mathrm{CW}}^{i}} \sum_{k} \left\| \pi(\mathbf{T}_{\mathrm{CW}}^{i}, \mathbf{P}_{\mathrm{W}}(k)) - \mathbf{p}^{i}(k) \right\|_2^2, \tag{7.4}$$

where $\mathbf{P}_{\mathtt{W}}(k)$ denotes the k-th 3D point in the map, $\mathbf{p}^i(k)$ denotes its corresponding feature in the i-th frame, and $\pi(\cdot)$ stands for the function of projecting a 3D point to its pixel location on the frame given the pose $\mathbf{T}^i_{\mathtt{CW}}$:

$$\pi(\mathbf{T}^i_{\mathtt{CW}}, \mathbf{P}_{\mathtt{W}}(k)) = \mathbf{K}\mathbf{R}^i_{\mathtt{CW}}\mathbf{P}_{\mathtt{W}}(k) + \mathbf{t}^i_{\mathtt{CW}}. \tag{7.5}$$

Once the pose of the frame is determined, VO triangulates 3D points in the newly observed areas, and the map gets updated in the same frequency as input.

Especially, in our implementation, edgeIS also identifies the 3D points belonging to moving objects and computes their poses individually. The process of pose estimation, as well as tracking observed objects, is shown in Fig. 7.5b. To compute the pose of the mobile device itself, edgeIS mainly selects 3D points which are labeled as background and their matched features to perform the bundle adjustment and get the current pose $\mathbf{T}^i_{\mathtt{CW}}$. Then for each object, edgeIS uses all 3D points belonging to it and the corresponding features to estimate the device's pose relative to the object, denoted as $\mathbf{T}^i_{\mathtt{CO}}$. If the object is moving, the estimated poses of the mobile device with two different reference systems will be different. With this difference, the object's motion relative to the background can be computed as

$$\mathbf{T}^i_{\mathtt{CO}} = \underset{\mathbf{T}^i_{\mathtt{CO}}}{\arg\min} \sum_k \left\| \pi\left(\mathbf{T}^i_{\mathtt{CO}}, \mathbf{Q}_{\mathtt{O}}(k)\right) - \mathbf{q}^i(k) \right\|_2^2. \tag{7.6}$$

To associate the dynamic object's pose in real-world reference, the following transformation is needed:

$$\begin{aligned} \mathbf{T}_{\mathtt{OW}} &= \mathbf{T}_{\mathtt{OC}}\mathbf{T}_{\mathtt{CW}} = \mathbf{T}^{-1}_{\mathtt{CO}}\mathbf{T}_{\mathtt{CW}} \\ &= \begin{bmatrix} \mathbf{R}^{\mathrm{T}}_{\mathtt{CO}}\mathbf{R}_{\mathtt{CW}} & \mathbf{R}^{\mathrm{T}}_{\mathtt{CO}}(\mathbf{t}_{\mathtt{CW}} - \mathbf{t}_{\mathtt{CO}}) \\ \mathbf{0}^{\mathrm{T}} & 1 \end{bmatrix}, \end{aligned} \tag{7.7}$$

where the $\mathbf{T}_{\mathtt{OW}}$ is the required pose of the dynamic object in the world coordinate.

Leveraging the annotation and grouping of the 3D points, the pose of each object gets to be updated individually and thus yields better performance in dynamic scenarios. Besides, it is worth mentioning that performing BA requires at least three pairs of 3D points and matched features, which is a small number. Therefore, if the number of points belonging to an object is less than 3, edgeIS takes it as either too small or too far away for accurate estimation.

7.3.3 *Mask Prediction*

The motion insights and 3D representations of observed environments provided by VO are crucial for anticipating the segmentation mask for the incoming frame, eliminating the necessity for deep learning methodologies. Upon the arrival of a new frame, edgeIS first deduces its orientation and enriches the environmental model accordingly. Nonetheless, accurately depicting each instance's mask within the frame poses a series of challenges.

The initial challenge involves identifying a collection of preceding frames that edgeIS relies on for predicting the current frame's mask. Through aligning feature pixels with pre-existing 3D points that are annotated, edgeIS acquires preliminary data on the objects present in the current scene. For every identified object, edgeIS seeks out frames that not only offer a clear view of the object (i.e., the object is entirely visible without being obscured) but also possess a viewpoint akin to the current frame (ensuring the angular disparity between frames is minimal). This procedure seamlessly aligns with the VO framework, as every 3D point retains data on its observational frames and the angular variance between poses is readily calculable.

Identifying the appropriate source frames leads to the subsequent inquiry of how to leverage their object masks to generate the mask for the current frame. Given the sparsity of features utilized for matching and mapping relative to the image resolution, directly reapplying the mask onto the new frame is not feasible. A fundamental notion is that an object mask's outline shapes it. Thus, pinpointing the edge pixels of an instance mask on the new frame effectively determines the mask's shape.

In edgeIS, we first extract the contour of the object mask on the source frame using the `findContours` function in OpenCV. The extracted contour, denoted as S, is represented by a list of connected pixels. Then for all $s_i \in S$, if not a feature pixel itself, edgeIS performs a search for k closest features within the mask. k is an empirical value based on our observation that the actual positions in 3D space corresponding to a small neighborhood of the object mask are not likely to experience shape changes in depth (in our implementation, we set k to 5). With the average depth as its own depth, s_i can be projected to the current frame with the relative pose and the projection function $\pi(\cdot)$. Applying the procedure to all observed objects and labeling other pixels as background, the mask for the current frame is generated by edgeIS for future use.

7.4 Contour Instructed Edge Inference Acceleration

In edge-enabled frameworks, despite processing urgent tasks on mobile units, the overall efficacy of such systems is significantly bound by the precision and speed of executing complex tasks at the edge [11, 22]. Distinguished from prior tasks centered on object detection, achieving instance segmentation on an edge server in a real-time manner with notable accuracy remains a formidable challenge. Typically, the duration required to produce instance masks constitutes a substantial fraction of the overall latency, thereby potentially degrading system performance due to the delay in receiving the image mask.

Within edgeIS, we introduce a *contour-focused edge inference acceleration* technique. By meticulously analyzing each component of RoI-centric segmentation deep learning architectures, we identified a method to expedite model inference by utilizing preliminary predictions of object borders to eliminate superfluous computations. Herein, we utilize Mask R-CNN [23]—a leading and highly precise model for instance segmentation—as an exemplar to delineate the adjustments applied to the network's structure. This discussion begins with an outline of the *dynamic anchor placement* strategy, illustrating how anticipated masks facilitate the reduction of nonessential operations, followed by an explanation of the *RoI pruning* approach, which enhances the process of pinpointing active regions.

7.4.1 Dynamic Anchor Placement

The architecture of Mask R-CNN consists of two stages. The first stage is Region Proposal Network (RPN) which proposes candidate object boxes for the following tasks. And in the second stage, tasks including classification, object detection, and mask prediction are executed in parallel. When taking an image as input without extra information, RPN needs to slide a small network across the whole convolutional feature map generated by the backbone network to propose initial bounding boxes. Calculating at every sliding window location in the feature map while the object only occupies a small portion of the image, this process inevitably produces a large number of redundant operations.

With the transferred mask of the object from the mobile device, the model can be "instructed" to perform more targeted region proposal, inspired by which we design the *dynamic anchor placement*. First, a surrounding box is calculated from the mask of each object. Then all convolutional layers in the backbone of RPN are registered with the size of feature maps they produced since the FPN (Feature Pyramid Backbone) can provide features from different levels. Apart from areas containing objects, new areas captured by the mobile device as mentioned in section Sect. 7.5 are also annotated with an initial box. Once the feature for region proposal is determined, each initial bounding box in the image will be downsampled to form a rectangular area in the feature map according to the scale from image size to feature size with an empirical padding. As illustrated in Fig. 7.6, to propose regions likely to contain objects, the range of the feature map that RPN needs to traverse and place anchors is limited within these areas. Furthermore, *dynamic anchor placement* rid the RPN of a large portion of unnecessary calculations by leveraging mobile side information.

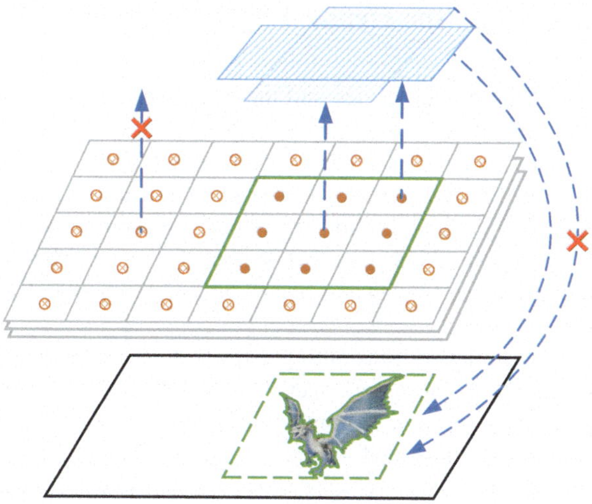

Fig. 7.6 Illustration of model acceleration. The computation of deep learning model is reduced by two functions performing on different stages, namely dynamic anchor placement and RoI pruning

7.4.2 RoI Pruning

At each sliding window location, several anchor boxes of different shapes are used to generate possible regions with box coordinates and class confidence. Though some selection processes are executed to reduce the number of RoIs used in the second stage, with the prior knowledge of the approximate location and the class of the object, this pruning process can be optimized for better efficiency and accuracy.

First, edgeIS groups all RoIs by the area annotated in the *dynamic anchor placement*. Then RoIs coming from the same area of a known object with class label c are sorted according to their confidence score on this class. In the sorted queue, an IoU score is computed between each RoI and the initial box of the object. (The computing metrics of IoU is described in Sect. 7.6.2.) As shown in Fig. 7.6, an RoI will be pruned if there exists an RoI with both higher confidence score on class c and IoU score with the initial box of the object. As for RoIs from the area corresponding to the unknown contents in the image, the Fast NMS introduced by Bolya et al. [8] is applied to perform the selection.

The intuition behind the pruning process is that for each discarded RoI, there is a better candidate for accurate mask prediction. By utilizing the prior information of the class and the estimated location of the object, *RoI pruning* saves the second stage where a mask is generated within each RoI from a large portion of unnecessary computations. Besides, the pruning of different classes can run in parallel, while IoU computing and batch sorting are already available in current frameworks. Therefore, the operations can be efficiently implemented and accelerated by standard GPU.

7.5 Content-Based Fine-Grained ROI Selection

The performance of the mobile device and the edge server is highly dependent on transmission [19]. Specifically, the transmission scheme should meet the requirements of both providing images (or some critical areas in images) of high resolution for the inference on the edge server and reducing transmission latency for real-time performance on the mobile side. In edgeIS, we propose a *content-based fine-grained RoI selection* scheme that first decides the frame to be transmitted and then efficiently compresses frames and masks on fine-grained RoI level based on the image content.

To determine the timing of transmitting a frame to the edge, edgeIS leverages the motion information obtained from Sect. 7.3.2. As the device moves, the content observed at the current location is continuously updated with the device's pose and modeled objects. As illustrated in Fig. 7.7b, if the matching proportion is less than a threshold t (in practice, we set t to 0.7), edgeIS will conduct pixel-level annotation. Apart from unseen areas, edgeIS checks the motion of observed objects, especially dynamic ones. If an object's pose changes significantly over a period, the transmission is also triggered for mask correction.

The frame to be transmitted is first divided into several different types of rectangular areas. As shown in Fig. 7.7c, for dynamic objects marked with red

(a) (b) (c) (d)

Fig. 7.7 An example of RoI selection module. (**a**) Segmentation masks before the scenario changes. (**b**) Frame in which the device and an object move. (**c**) Area partition. (**d**) Encoded frame with blue and yellow areas compressed features and unknown areas with green ones in Fig. 7.7b, the corresponding areas are assigned with low compression levels to preserve the high quality of the image. The remaining blocks, containing either background pixels or static objects (marked with yellow and blue features correspondingly), are then compressed to low quality without harming the performance of the segmentation model on the edge server. Moreover, detailed information such as coordinates and class labels of each block will also be packed to help decode and process the frame on the edge. We use the definition of tile defined in HEVC [24] and implement the encoding method based on Kvazaar [25], which encodes each tile in a frame individually and therefore is suitable for parallel acceleration.

As for the mask generated on the edge, instead of transmitting the whole mask back to the mobile device, edgeIS only extracts and sends the contour and class label of each instance back to the mobile device, which are enough to reproduce and render the whole mask. By compressing individual blocks of the frame differently according to their contents, our transmission module achieves the goal of reducing the amount of transmitted data while maintaining high accuracy. By using pixel coordinates rather than image representation, the transmission cost of sending back masks is also considerably reduced.

7.6 Evaluation for EdgeSLAM-Based Environment Understanding

7.6.1 Implementation

Client We implement the mobile part of edgeIS on mobile devices, including an Apple iPhone 11 and a Samsung Galaxy 10. The functions of the mobile module are mainly implemented in C++ for ease of cross-platform deployment. For input and output, we use OpenCV [26] for feeding video frames at fixed 30 fps and rendering masks and visual effects on the screen. The core parts of the mask transfer module follow the workflow in Fig. 7.4, and we modify the VO from ORB-SLAM [27] to realize motion tracking and object modeling, based on which the functions of object tracking and mask prediction are implemented. Besides, an additional thread is used for communicating with the transmission module to send frames according to the mobile device and receive masks from the server.

Server We deploy our edge-side module on an Nvidia Jetson TX2 edge server. For the deep learning model, we choose Mask R-CNN [23] with a ResNet-101-FPN as backbone feature extractor, and the parameters are pretrained on COCO image dataset [28]. Our *contour instructed edge inference acceleration*, as well as the Mask R-CNN, is implemented in Python 3.6.5 with PyTorch 0.4.0 [29].

Remote Data Interaction As introduced in Sect. 7.5, we implement our video encoder on the mobile side based on Kvazaar [25] in C++, and the decoder on the server side is also implemented in C++ with the OpenHEVC [30] library. For information such as vertices of the contour, we use C++ Boost [31] library for the serialization and transmission between the two sides.

7.6.2 Experiment Setup

We perform an extensive experimental validation of edgeIS on two public video datasets, namely DAVIS [32] and Xiph [33], as well as a handcrafted dataset with typical indoor and outdoor AR scenarios. The videos we select for processing from the two public datasets contain a total number of 6534 frames, and the videos in our dataset have a total lasting time of 7.3 min with 13,276 frames. In our experiments, all videos are set to an input rate of 30 fps with uniform resolution.

Similar to previous works, the most critical performance we care about an edge-assisted system is the accuracy, since latency and other factors will all lead to a delayed rendering and finally a decrease in accuracy [12, 19]. In practice, we use the average IoU that represents the similarity between predictions and ground truths to measure the accuracy of edgeIS for two individual tasks. For an object, denoting the set of pixels in its ground truth mask as $S_{groundtruth}$ and the set of pixels in the predicted mask as $S_{prediction}$, IoU is given by

$$\mathbf{IoU} = \frac{|S_{groundtruth} \cap S_{prediction}|}{|S_{groundtruth} \cup S_{prediction}|}. \tag{7.8}$$

A high IoU score indicates that the predicted mask accurately covers the ground truth without taking in too many unrelated pixels.

To evaluate the segmentation performance, we implement two baseline approaches: running entirely on the mobile device and on the edge node. The former one uses TensorFlow Lite [34] to fully deploy the DL model on the mobile devices. The latter one sends all frames to the edge, performing segmentation and sending results back to the mobile device with a best-effort strategy. Besides, we apply a motion vector-based scheme on the mobile side to approximately track the mask locally.

Additionally, we modify two previous systems, EdgeDuet [10] and EAAR [11], designed for real-time object detection with edge-assisted architecture to perform the instance segmentation task. We adopt their transmission strategy and replace the object detection model on the edge server with the same Mask R-CNN as edgeIS uses. On the mobile side, we adopt the local tracker proposed by them based on either KCF or motion vector to update the contour of observed objects rather than detection boxes.

7.6.3 Performance Comparison

We first evaluate the instance segmentation accuracy of edgeIS and compared systems. In our evaluation, we adopt a loose threshold of 0.5 and a strict threshold of 0.75 which are commonly used in computer vision community to determine an acceptable accuracy. IoU smaller than the threshold is called a false result. We mainly focus on the strict threshold since high accuracy is required for target application scenarios.

7.6.3.1 Overall Performance

Figure 7.8 illustrates the CDF of instance segmentation accuracy. The false segmentation rate of pure mobile computing and best-effort edge assistance is 78.3% and 60.1%, respectively. With local tracker and transmission schemes, *EAAR* and *EdgeDuet* reduce the false rate to 21% and 39%, respectively, outperforming these straight-forward methods. However, the trackers designed for objection detection are still too coarse for segmentation task, and the segmentation model fails to provide timely results without acceleration. With motion information obtained for more accurate mask prediction, the false rate of edgeIS is only 3.9%, reducing that of *EAAR* and *EdgeDuet* by 82% and 91% correspondingly. Besides, the average accuracy of edgeIS is 0.92, improving that of the two systems by 10% and 20%, respectively.

7.6.3.2 Performance under Different Networks

Since the network conditions vary in real scenarios and have an important influence on edge-assisted systems, we further evaluate the performance under different network settings including Wi-Fi 2.4 GHz and Wi-Fi 5 GHz. We first evaluate the segmentation accuracy of edgeIS. As shown in Fig. 7.9, for segmentation tasks, the false rate of edgeIS under Wi-Fi 2.4 GHz and Wi-Fi 5 GHz is 6.1% and 4.1%, respectively, while the false rate of *EAAR* and *EdgeDuet* under the fastest Wi-Fi 5 GHz is 21% and 41%, which will be higher when switched to Wi-Fi 2.4 GHz. Under either network condition, edgeIS reduces the false rate by at least 78%

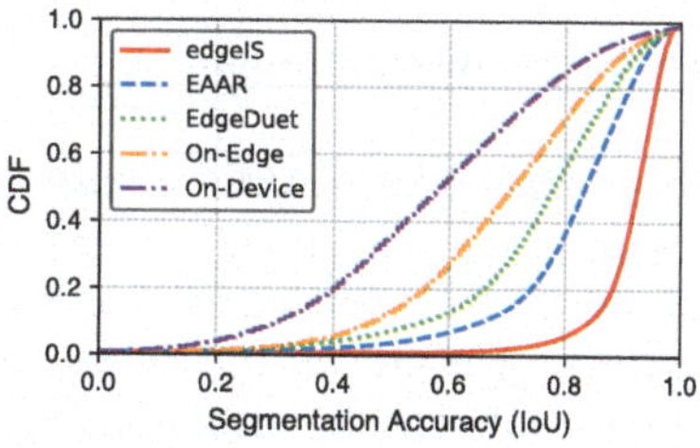

Fig. 7.8 Overall performance comparison

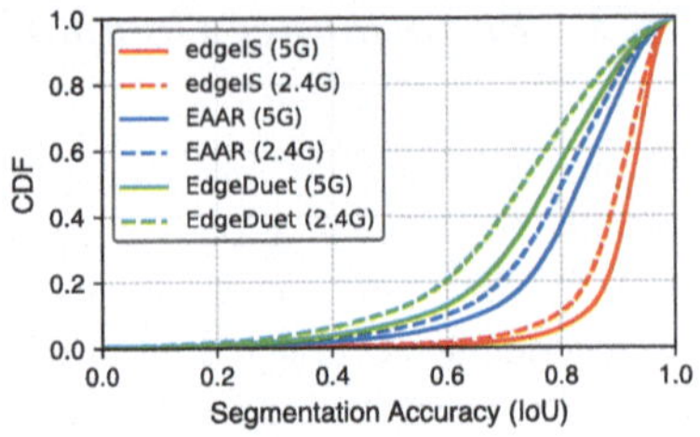

Fig. 7.9 Performance with network conditions

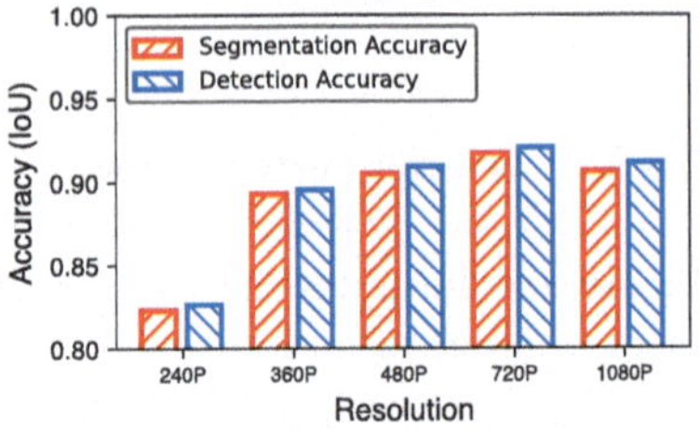

Fig. 7.10 Impact of image resolutions

compared with *EAAR* and 83% compared with *EdgeDuet*, proving effective in various circumstances.

7.6.4 Robustness Evaluation

7.6.4.1 Video Resolution

Figure 7.10 shows the accuracy performance of edgeIS with videos of different resolutions as input. With various video settings, edgeIS achieves a segmentation accuracy higher than 0.81 and a detection accuracy higher than 0.83. One interesting finding is that the best performance is not achieved with the input of the highest resolution. The increase in resolution brings a trade-off phenomenon between quality of feature extraction and computational cost. Features of higher quality benefit the motion estimation and object modeling, while a higher cost on processing input frames incurs longer latency which further impacts on accuracy. In our experiments, both the best segmentation accuracy of 0.91 and the best detection accuracy of 0.92 are achieved with input of 720P.

7.6.4.2 Camera Motion

To evaluate edgeSLAM's robustness against different moving status, we record videos of the same route with people walking, striding, and jogging. Figure 7.11 shows that the false rate of edgeIS is 4.7%, 9.8%, and 29.9%, respectively, in slow, medium, and fast circumstances. In the worst case, edgeIS can still achieve an average IoU of 0.82. The results demonstrate that with motion information used

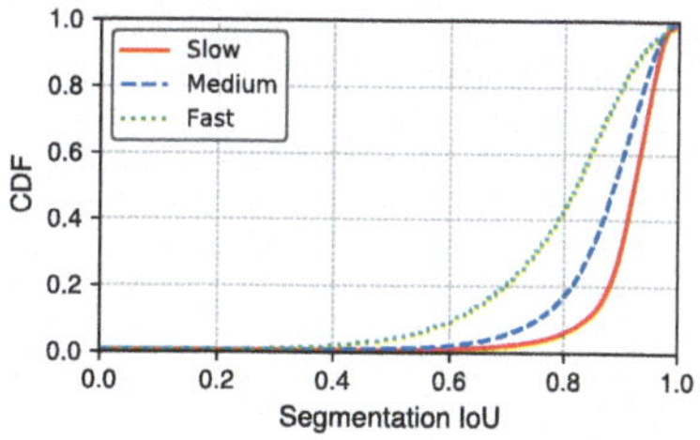

Fig. 7.11 Impact of movement speeds

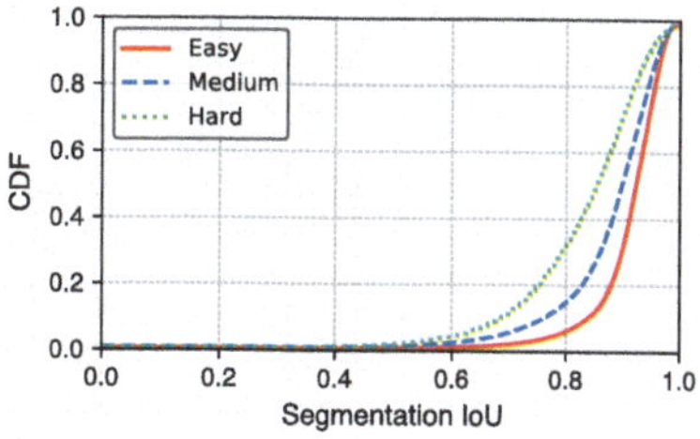

Fig. 7.12 Impact of scene complexities

for predicting masks and transmitting data, edgeIS proves robust against different moving speed and competent for most application scenarios.

7.6.4.3 Scene Complexity

To evaluate edgeSLAM's performance in various scenarios, we manually arrange scenes of different complexity. The number of objects in simple scenes is no more than 3 and rises to 10 in medium scenes. In scenes of hard level, objects will move during the experiment. As shown in Fig. 7.12, the average accuracy in terms of IoU in easy, medium, and complex scenes is 0.91, 0.88, and 0.83, respectively. In scenes where objects move during the process, the false rate of edgeIS is 19.7%, indicating that most of the observed objects can be well segmented. The results prove that tracking objects' 3D poses effectively improves the robustness of edgeIS in dynamic environments, making it applicable in real scenarios.

7.6.5 Benefits of Individual Components

We further experimentally analyze the three core components of edgeIS, particularly the performance gains that each of them brings into the overall system. In the following, we abbreviate the three modules, ***m**otion **a**ware mobile **m**ask **t**ransfer*, ***c**ontour **i**nstructed edge **i**nference **a**cceleration*, and ***c**ontent-based **f**ine-grained **R**oI **s**election* as MAMT, CIIA, and CFRS, respectively.

Fig. 7.13 Effectiveness of each component

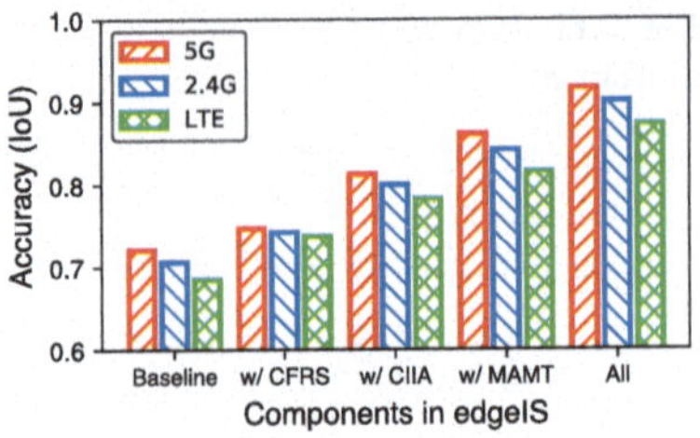

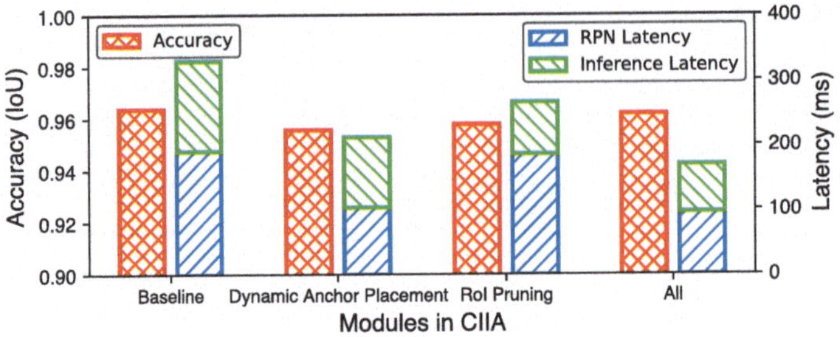

Fig. 7.14 Effectiveness of model acceleration

7.6.5.1 Benefits of Individual Modules

In this part, we analyze the improvements obtained with the three core components in edgeIS individually. The baseline method we use is the best-effort edge assistance with motion vector tracking mentioned in Sect. 7.6.2. As shown in Fig. 7.13, with CFRS to reduce transmitted data, edgeIS achieves a more even performance under different network conditions, and the accuracy is improved by 3–7% due to reduced transmission latency. With CIIA, the inference latency on the edge is significantly reduced and thus provides more timely results. The accuracy is improved by 12–14% under different network connections. The improvement gained with MAMT is more than 19% since motion information directly benefits mask prediction on the mobile side. Compared with baseline methods, edgeIS achieves an overall improvement of 27% on accuracy under all network conditions.

7.6.5.2 Benefits of Model Acceleration

We further examine the latency improvement that model acceleration brings to Mask R-CNN. The overall latency is divided into RPN latency and inference latency. As in Fig. 7.14, with dynamic anchor placement, RPN latency is reduced by 46% benefiting from discarding unnecessary anchor computation, and inference latency is reduced by 21% with less RoI produced. With RoI pruning, the inference latency is reduced by 43%. Overall, the model acceleration module reduces the latency by 48% while maintaining an accuracy higher than 0.92.

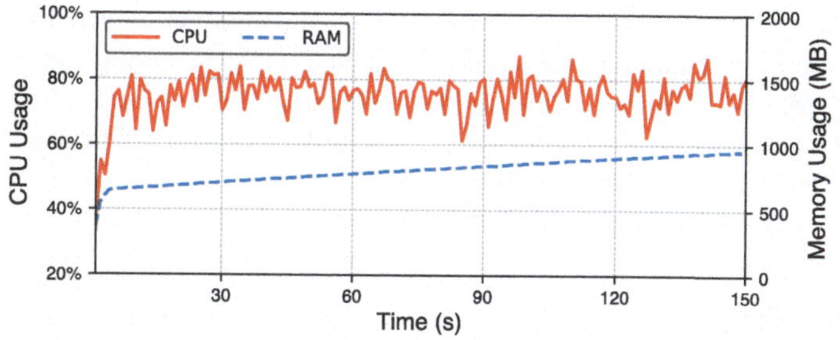

Fig. 7.15 CPU and memory usages on device

7.6.6 Resource Overhead

7.6.6.1 Mobile Resource Usage

As depicted in Fig. 7.15, we measure the CPU and memory usage of edgeIS for a period of time after initialization on iPhone 11. The CPU usage of the system is around 75%, and the memory usage is increased by nearly 2 MB/s. The increase in memory usage comes from the recording of new data in picture frames and local maps. Through the additional clearing algorithm, the system can periodically clear the data of low utilization to control the global memory usage within 1 GB, allowing edgeIS to run smoothly on the latest mobile phones.

7.6.6.2 Power Consumption

To measure power consumption, we start with full power on iPhone 11 and Samsung Galaxy 10 and record the power consumption within 10 min through the management application. The experiment is repeated dozens of times. On average, the power consumption of edgeIS in 10 min is 4.2% and 5.4%, respectively, on iPhone11 and Galaxy 10, which is comparable to running ARKit and ARCore demo applications continuously for the same time duration.

7.7 Related Work

Instance Segmentation Deep learning (DL) models have been well studied in recent years to perform accurate instance segmentation tasks. Typical methods usually follow the routine of generating pixel-level segmentation results within a proposed region [35, 36]. Mask R-CNN [23] adds a branch based on Faster R-CNN [37] to perform segmentation in parallel with object detection. Pioneer studies such as PANet [38] and FCIS [36] also make efforts to better exploit information inside the detection box to improve segmentation accuracy. Apart from these phased solutions, there emerge end-to-end segmentation models that are free of region proposal schemes using techniques such as pixel clustering and coefficient regression [8, 39]. However, these models are usually less accurate than region-based ones and hard to decomposite, leaving little room for improvement when deployed for mobile applications. In edgeIS, we take full advantage of the region-based model's high accuracy and significantly reduce its latency using mobile side information.

Edge Offloading Offloading computational intensive tasks to powerful edge servers to enable complex mobile applications has been a popular research topic. The most important concern about the edge-assisted architecture is the balance between the two sides and latency incurred by offloading [19]. In recent years, the "detect+track" framework that only requires key information from the edge server and performs tracking locally on mobile devices has proved effective. Some methods [12, 40, 41] design mechanisms to decide whether to upload the content to the server or update the result locally. Chameleon [42] and VideoStorm [43] apply video configurations to achieve higher accuracy with the same amount of resources on the server. Besides, some previous systems focus on reducing transmission latency. The recent work of Liu et al. [11] introduces an edge-assisted system achieving high accuracy object detection and human keypoint detection task on existing AR/MR systems running at 60 fps for both the object detection. EdgeDuet [10] further decomposites the offloading pipeline to tile level for more efficient video transmission. While existing techniques mainly use the edge server to assist mobile devices, edgeIS leverages information unique on the mobile side to build a two-way instructional system to achieve better performance in our "transfer+infer" paradigm.

Visual Odometry Visual odometry takes in consecutive frames to estimate device trajectory and construct a 3D map of environment [44–46]. Pioneer studies apply filtering-based approaches [47], which are replaced by optimization-based ones [48] for better accuracy and efficiency. ORB-SLAM [49], one of the most representative works, uses ORB features for tracking and mapping and achieves excellent performance. Using visual clues, visual odometry has the potential of combining with other vision-based methods to realize applications, such as indoor navigation and robot collaboration [50–52]. edgeIS exploits the sensing ability of visual odometry and integrates it with accurate instance annotations to enable robust object tracking and segmentation on mobile devices.

7.8 Conclusion

In this chapter, we propose edgeIS, an instance segmentation solution for mobile devices, achieving both accuracy and real-time simultaneously with the assistance of edge computation resources. Distinguished from existing "track+detect" edge-assisted architecture, we redesign and propose a fresh "transfer+infer" mobile-edge collaboration paradigm which achieves a two-way promotion of both sides by fully exploiting the motion and environmental sensing capabilities of mobile devices. We fully implement edgeIS on a resource-limited edge node and different types of mobile devices and extensively evaluate the performance on three datasets under different network conditions. The experimental results and a field study in an industrial scenario show that edgeIS achieves satisfying results in all scenarios.

References

1. A.M. Hafiz, G.M. Bhat, A survey on instance segmentation: state of the art. Int. J. Multimed. Inf. Retr. **9**(3), 1–19 (2020)
2. M. Hololens, https://www.microsoft.com/en-us/hololens
3. F. Ahmad, H. Qiu, R. Eells, F. Bai, R. Govindan, CarMap: fast 3D feature map updates for automobiles, in *Proceedings of the USENIX NSDI* (2020)
4. K. Kanistras, G. Martins, M.J. Rutherford, K.P. Valavanis, A survey of unmanned aerial vehicles (UAVs) for traffic monitoring, in *2013 International Conference on Unmanned Aircraft Systems (ICUAS)* (IEEE, New York, 2013), pp. 221–234
5. J. Xu, G. Chi, Z. Yang, et al., FollowUpAR: enabling follow-up effects in mobile AR applications, in *Proceedings of the ACM MobiSys* (2021)
6. S. Minaee, Y.Y. Boykov, F. Porikli, A.J. Plaza, N. Kehtarnavaz, D. Terzopoulos, Image segmentation using deep learning: A survey. IEEE Trans. Pattern Anal. Mach. Intell. **44**(7), 3523–3542 (2021)
7. M. Gamal, M. Siam, M. Abdel-Razek, ShuffleSeg: real-time semantic segmentation network. arXiv preprint arXiv:1803.03816 (2018)
8. D. Bolya, C. Zhou, F. Xiao, Y.J. Lee, YOLACT: real-time instance segmentation, in *Proceedings of the IEEE/CVF International Conference on Computer Vision* (2019), pp. 9157–9166
9. S. Yi, Z. Hao, Q. Zhang, Q. Zhang, W. Shi, Q. Li, LAVEA: Latency-aware video analytics on edge computing platform, in *Proceedings of the Second ACM/IEEE Symposium on Edge Computing* (2017), pp. 1–13
10. X. Wang, Z. Yang, J. Wu, Y. Zhao, Z. Zhou, EdgeDuet: tiling small object detection for edge assisted autonomous mobile vision, in *Proceedings of the IEEE INFOCOM* (2021)
11. L. Liu, H. Li, M. Gruteser, Edge assisted real-time object detection for mobile augmented reality, in *Proceedings of the ACM MobiCom* (2019)
12. T.Y.-H. Chen, L. Ravindranath, S. Deng, P. Bahl, H. Balakrishnan, Glimpse: continuous, real-time object recognition on mobile devices, in *Proceedings of the ACM Sensys* (2015)
13. J. Xu, H. Cao, D. Li, et al., Edge assisted mobile semantic visual slam, in *IEEE INFOCOM 2020-IEEE Conference on Computer Communications* (IEEE, New York, 2020), pp. 1828–1837
14. *Nvidia jetson tx2*. https://developer.nvidia.com/embedded/jetson-tx2
15. *Nvidia Jetson AGX Xavier*. https://developer.nvidia.com/embedded/jetson-agx-xavier
16. D. Nistér, O. Naroditsky, J. Bergen, Visual odometry, in *Proceedings of the 2004 IEEE Computer Society Conference on Computer Vision and Pattern Recognition, 2004. CVPR 2004*, vol. 1 (IEEE, New York, 2004), pp. I–I
17. T. Qin, P. Li, S. Shen, VINS-Mono: a robust and versatile monocular visual-inertial state estimator. IEEE Trans. Robot. **34**(4), 1004–1020 (2018)
18. R. Mur-Artal, J.D. Tardós, ORB-SLAM2: an open-source SLAM system for monocular, stereo and RGB-D cameras. IEEE Trans. Robot. **33**(5), 1255–1262 (2017)
19. Z. Chen, W. Hu, J. Wang, et al., An empirical study of latency in an emerging class of edge computing applications for wearable cognitive assistance, in *Proceedings of the IEEE/ACM SEC* (2017)

20. E. Rublee, V. Rabaud, K. Konolige, G. Bradski, ORB: an efficient alternative to sift or surf, in *Proceedings of the 2011 International Conference on Computer Vision (ICCV)* (IEEE, New York, 2011), pp. 2564–2571
21. R. Hartley, A. Zisserman, *Multiple View Geometry in Computer Vision* (Cambridge University, Cambridge, 2003)
22. J. Yi, S. Choi, Y. Lee, EagleEye: wearable camera-based person identification in crowded urban spaces, in *Proceedings of the 26th Annual International Conference on Mobile Computing and Networking* (2020), pp. 1–14
23. K. He, G. Gkioxari, P. Dollár, R. Girshick, Mask R-CNN, in *Proceedings of the IEEE ICCV* (2017)
24. K. Misra, A. Segall, M. Horowitz, S. Xu, A. Fuldseth, M. Zhou, An overview of tiles in HEVC. IEEE J. Sel. Top. Sign. Proces. **7**(6), 969–977 (2013)
25. *Kvazaar*. https://github.com/ultravideo/kvazaar
26. G. Bradski, The OpenCV Library. Dr. Dobb's J. Softw. Tools **25**(11), 120–123 (2000)
27. *Orb-slam2*. https://github.com/raulmur/ORB_SLAM2 (2016)
28. T.-Y. Lin, M. Maire, S. Belongie, et al., Microsoft COCO: common objects in context, in *European Conference on Computer Vision* (Springer, Berlin, 2014)
29. *PyTorch mask R-CNN*, https://github.com/multimodallearning/pytorch-mask-rcnn (2017)
30. *OpenHEVC*, https://github.com/OpenHEVC/openHEVC
31. S. Koranne, Boost c++ libraries, in *Handbook of Open Source Tools* (Springer, Berlin, 2011), pp. 127–143
32. S. Caelles, J. Pont-Tuset, F. Perazzi, A. Montes, K.-K. Maninis, and L. Van Gool, "The 2019 Davis challenge on VOS: Unsupervised multi-object segmentation," *arXiv:1905.00737*, 2019.
33. C. Montgomery, H. Lars, Xiph.org video test media (Derf's collection). https://media.xiph.org/video/derf (1994)
34. *TensorFlow lite*. https://tensorflow.org/lite
35. J. Dai, K. He, Y. Li, S. Ren, J. Sun, Instance-sensitive fully convolutional networks, in *Proceedings of ECCV* (Springer, Berlin, 2016)
36. Y. Li, H. Qi, J. Dai, X. Ji, Y. Wei, Fully convolutional instance-aware semantic segmentation, in *Proceedings of the IEEE CVPR* (2017)
37. S. Ren, K. He, R. Girshick, J. Sun, Faster R-CNN: towards real-time object detection with region proposal networks, in *Advances in Neural Information Processing Systems* (2015)
38. S. Liu, L. Qi, H. Qin, J. Shi, J. Jia, Path aggregation network for instance segmentation, in *2018 IEEE/CVF Conference on Computer Vision and Pattern Recognition* (2018), pp. 8759–8768. https://doi.org/10.1109/CVPR.2018.00913
39. D. Neven, B.D. Brabandere, M. Proesmans, L.V. Gool, Instance segmentation by jointly optimizing spatial embeddings and clustering bandwidth, in *Proceedings of the IEEE/CVF Conference on Computer Vision and Pattern Recognition* (2019), pp. 8837–8845
40. K. Chen, T. Li, H.-S. Kim, D.E. Culler, R.H. Katz, Marvel: enabling mobile augmented reality with low energy and low latency, in *Proceedings of the 16th ACM Conference on Embedded Networked Sensor Systems* (2018), pp. 292–304
41. X. Ran, H. Chen, X. Zhu, Z. Liu, J. Chen, DeepDecision: a mobile deep learning framework for edge video analytics, in *Proceedings of the IEEE INFOCOM* (2018)
42. J. Jiang, G. Ananthanarayanan, P. Bodik, S. Sen, I. Stoica, Chameleon: scalable adaptation of video analytics, in *Proceedings of the ACM SIGCOMM* (2018)
43. H. Zhang, G. Ananthanarayanan, P. Bodik, M. Philipose, P. Bahl, M.J. Freedman, Live video analytics at scale with approximation and delay-tolerance, in *Proceedings of the USENIX NSDI* (2017)
44. A.J. Davison, Real-time simultaneous localisation and mapping with a single camera, in *Proceedings of the IEEE ICCV* (2003)
45. C. Forster, M. Pizzoli, D. Scaramuzza, SVO: fast semi-direct monocular visual odometry, in *2014 IEEE International Conference on Robotics and Automation (ICRA)* (IEEE, New York, 2014), pp. 15–22

46. J. Xu, H. Chen, K. Qian, et al., IVR: integrated vision and radio localization with zero human effort, in *PACM on Interactive, Mobile, Wearable and Ubiquitous Technologies* (2019)
47. A.J. Davison, I.D. Reid, N.D. Molton, O. Stasse, MonoSLAM: Real-time single camera slam. IEEE Trans. Pattern Anal. Mach. Intell. **29**(6), 1052–1067 (2007)
48. G. Klein, D. Murray, Parallel tracking and mapping for small AR workspaces, in *Proceedings of the IEEE ISMAR* (2007)
49. R. Mur-Artal, J.M.M. Montiel, J.D. Tardos, ORB-SLAM: a versatile and accurate monocular slam system' IEEE Trans. Robot. **31**(5), 1147–1163 (2015)
50. E. Dong, J. Xu, C. Wu, Z. Yang, Y. Liu, Pair-Navi: peer to peer indoor navigation with mobile visual slam, in *Proceedings of the IEEE INFOCOM* (2019)
51. G. Chi, J. Xu, J. Zhang, Q. Zhang, Q. Ma, Z. Yang, Locate, tell, and guide: enabling public cameras to navigate the public. IEEE Trans. Mob. Comput. **22**(2), 1010–1024 (2021)
52. P. Schmuck, M. Chli, CCM-SLAM: robust and efficient centralized collaborative monocular simultaneous localization and mapping for robotic teams. J. Field Rob. **36**(4), 763–781 (2019)

Chapter 8
Multi-user AR with EdgeSLAM

Abstract Most existing multi-user augmented reality (AR) systems only support multiple colocated users to view a common set of virtual objects but lack the ability to enable each user to directly interact with other users appearing in his/her view. Such multi-user AR systems should be able to detect the human keypoints and estimate device poses (for identifying different users) in the meanwhile. However, due to the stringent low latency requirements and the intensive computation of the above two capabilities, previous research only enables either of the two capabilities for mobile devices even with the aid of the edge server. Integrating the above two capabilities is promising but nontrivial in terms of latency, accuracy, and matching. To fill this gap, we propose DiTing to achieve real-time ID-aware multi-device visual interaction for multi-user AR applications, which contains three key innovations: *Shared On-device Tracking* to merge the similar computation for optimized latency, *Tightly Coupled Dual Pipeline* to enhance the accuracy of each task through mutual assistance, and *Body Affinity Particle Filter* to precisely match device poses with human bodies. We implement DiTing on four types of mobile AR devices and develop a multi-user AR game as a case study. Extensive experiments show that DiTing can provide high-quality human keypoint detection and pose estimation in real-time (30 fps) for ID-aware multi-device interaction and outperform the SOTA baseline approaches.

8.1 Introduction

Augmented reality (AR) has recently exploded in popularity among mobile users by providing innovative immersive interactive experiences. With the increasing capability and popularity of mobile devices, the multi-user augmented reality has boomed in diverse applications, including entertainment [1], education [2], industry training [3], etc. Most existing multi-user AR systems only provide the functionality that multiple colocated user's view a common set of virtual objects. For example, Pokémon Go permits multiple users to view their virtual creatures together in the same real world [4], and Minecraft allows multiple users to build structures together from virtual blocks [5]. However, a truly practical multi-user AR system

J. Xu et al., *Edge Assisted Mobile Visual SLAM*,
https://doi.org/10.1007/978-981-97-3573-0_8

should also enable multiple users to interact with one another directly, which means users can detect the humans in the view and distinguish their identities (IDs) for targeted interaction. For example, in a first-person shooting multi-user VR game, the user shoots other users appearing on his/her screen and causes damage according to which user he/she hits [6]. Schools or factories adopt multi-user AR systems to educate students or train employees in virtual scenarios, where the students or employees need to distinguish different people and cooperate with particular partners [7, 8]. Consequently, the multi-user AR system supporting ID-aware multi-device visual interaction is more desirable and promising than the traditional multi-user AR system which only supports multiple users to view common virtual objects.

In a nutshell, the multi-user AR system supporting ID-aware multi-device visual interaction should own two capabilities. The first is *detecting human keypoints* in real-time which allows rendering virtual overlays on the human bodies, such as a virtual weapon in the hand or a virtual identity tag above the head. The second is *distinguishing the identities* of all users on the screen so as to selectively interact with someone. Most existing research has focused on improving either of the two capabilities for mobile devices but does not consider supporting two capabilities in the meanwhile. As for the first capability, it could be enabled with CNN which has shown superior performance in the human keypoint detection task [9]. Some research has focused on enabling edge-assisted real-time video analysis on mobile devices, and the representative work EAAR [10] can achieve high accuracy human keypoint detection for the AR system at 60 fps by offloading parts of computations to the edge. Nevertheless, as illustrated in Fig. 8.1a, these researches fail to locate multiple users and distinguish their identities when detecting human keypoints. The second capability is essentially enabled with multi-device pose sharing, which is implemented by real-time pose estimation via visual odometry (VO). Existing VO methods can attain relative accurate pose estimation [11–13] with periodically updated device pose computed by backend relocalization, but running these VO systems directly on resource-constrained mobile devices will induce obvious pose estimation error (Fig. 8.1b). MIRSA [14] further designs a lightweight VO network for pose estimation and achieves real-time multi-device interaction for mobile devices. However, as shown in Fig. 8.1c, the mobile devices still cannot detect human keypoints and are unable to accurately render virtual overlays on human bodies. The most relevant work is edgeSLAM [15] which seemingly can run semantic segmentation and SLAM tasks meanwhile, but in fact edgeSLAM does not run semantic segmentation in real-time. Consequently, it is incapable of matching different user bodies with their device poses (Fig. 8.1d).

The above research gap motivates us to design and implement DiTing, an edge-assisted real-time ID-aware multi-device interaction framework for mobile multi-user augmented reality applications. Benefiting from the increasing AI processing capability of mobile devices and edge, it is promising to achieve both the human keypoint detection task and pose estimation task for mobile devices with the aid of the edge. Our vision is to fully aggregate the above two tasks which effectively reduces the frequent data transmission between the mobile device and the edge

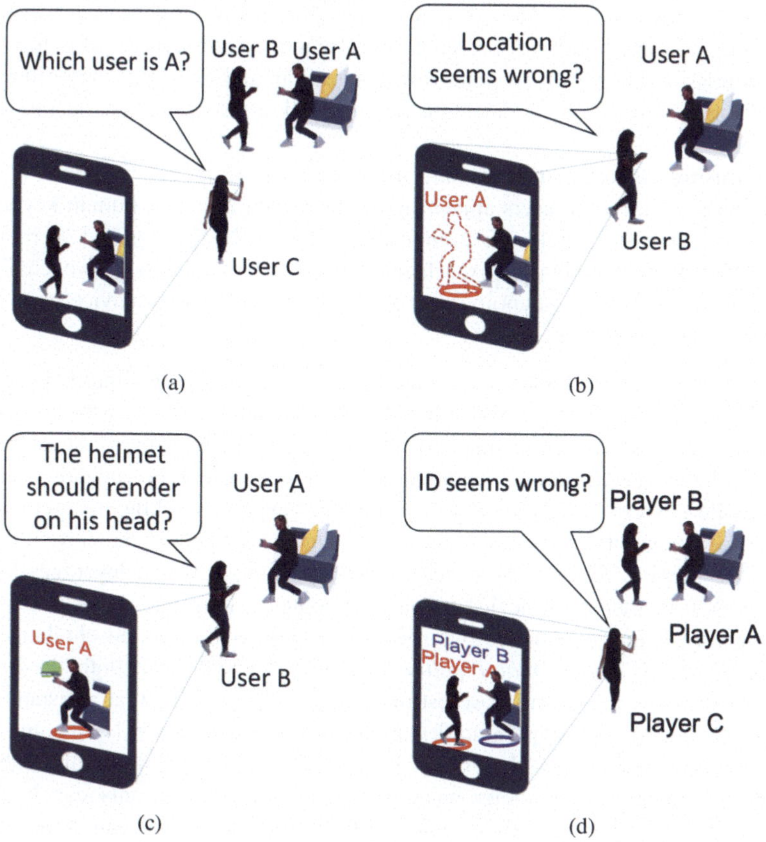

Fig. 8.1 The limitations of visual interaction in multi-user AR applications. (First-person shooting AR game as an example). (**a**) Lack of user ID information. (**b**) Location error in rendering. (**c**) Lack of human keypoint information. (**d**) Mismatch of user IDs

and trims redundant computation. Nevertheless, realizing the design of DiTing is nontrivial and faces three key challenges:

- **Real-time constraint of two on-device tasks**: Both the human keypoint detection task and pose estimation task are computationally intensive; thus they will induce hundreds of milliseconds computing latency on mobile devices. It makes the results out of date once produced in latency-sensitive AR applications running in real-time (e.g., 30 fps). Although offloading intensive computation to the edge enables a single task to meet the real-time requirement, it is still a challenge to run two tasks concurrently in real-time on mobile devices, which is a prerequisite for our vision.
- **Multi-user offloading competition for bandwidth**: Multiple users offloading video stream will cause bandwidth competition and lengthen transmission latency, degrading interaction experience when more users are involved in multi-

user AR applications. However, selectively offloading key frames seems an accuracy–latency trade-off problem: high-frequency offloading will induce long transmission latency or even network congestion, and low-frequency offloading will skip some frames' detection and expand pose estimation accumulative errors.

- **Confusing ID association among multiple users**: Suppose we have obtained the poses of multiple users appearing on the screen and their human keypoints. The former are 6-DOF localization results including 3D location and 3D rotation, and the latter are annotated pixels located on the 2D image. It is a novel challenge to accurately pair the keypoints on the user body with his/her device pose when the number of users increases.

To address the first challenge, we carefully divide two task pipelines between the mobile devices and the edge. We offload key frames to the edge for getting accurate human keypoint detection results and relocalization results of the pose estimation task. With the updated human keypoint detection results and relocalization results, the mobile devices simply need to run human keypoint tracking for getting the human keypoints of the current frame and to run VO for getting the current device pose. We propose *Shared On-device Tracking* technique which leverages hybrid optical flow to merge on-device human keypoint tracking and on-device VO to guarantee real-time experience over 30fps. To address the second challenge, we utilize *Tightly Coupled Dual Pipeline* method to benefit from both the human keypoint detection task and pose estimation tasks through mutual assistance. To be specific, the human keypoint information can help on-device VO overcome environment dynamics, reduce the essential offloading frequency, and finally enhance the performance. On the other hand, the pose estimation results can help the user foresee whether new users enter his/her view, so he/she can only offload those frames containing new appeared users to the edge. It effectively reduces transmission pressure on network without unnecessary offloading. To address the third challenge, we first transform 6-DOF device poses into 2D locations on the image and then employ *Body Affinity Particle Filter* technique to ensure the device poses are matched precisely with each human body. The key contributions of DiTing can be summarized as follows:

- We design DiTing enabling higher quality multi-user AR interaction, and to the best of our knowledge, it is the first edge-assisted multi-user AR system to achieve real-time ID-aware visual interaction.
- We conduct substantial preliminary experiments to measure the latency and accuracy of running either device pose estimation or human keypoint detection on the mobile device and the edge server, and we quantitatively analyze the challenges to integrate above two tasks on the mobile device.
- We propose *Shared On-device Tracking* technique to reduce nearly half computing latency on mobile devices, which maintains real-time experience over 30 fps.
- We develop *Tightly Coupled Dual Pipeline* method to improve the accuracy of device pose estimation by 92% and reduce bandwidth consumption by 60.3%.

- We create *Body Affinity Particle Filter* technique to settle the mismatch of user IDs when multiple users appear in the video.
- We implement DiTing on four different commodity mobile devices and develop a multi-user first-person shooting AR game as case studying. The results show DiTing achieves delightful results, outperforming all SOTA approach.

The rest of chapter is organized as follows. We analyze the challenges and introduce a basic solution in Sect. 8.2. We present an overview of DiTing in Sect. 8.3. The following three sections detailedly explain our three designed techniques separately. We introduce the detailed implementation of DiTing in Sect. 8.7 and evaluate the performance in Sect. 8.8. We review the related works in Sect. 8.9 and conclude this chapter in Sect. 8.10.

8.2 Challenges and Analysis

In this section, the challenges to integrate human keypoint detection task and pose estimation task on the mobile device in terms of latency, accuracy, and ID matching are analyzed quantitatively.

Latency Analysis Simultaneously accomplishing both the human keypoint detection and device pose estimation tasks on mobile AR devices is a multi-faceted challenge because the dual tasks are both too computationally intensive to be executed on-device or too bandwidth intensive to be simply offloaded to the edge server. As shown in Fig. 8.3a, the lightweight human keypoint detection model requires more than 120 ms on current high-end smartphones (tested on Galaxy S10 and Google Pixel 3) and more than 100 ms on Nvidia Jetson TX2 which is a powerful development board and often regarded as the edge. As to the device pose estimation, the VO module optimized for the mobile device requires about 23 ms on smartphones and suffers from accumulative error, which causes spatial inconsistency during long-term pose tracking. Thus relocalization is essential for accurate pose estimation, as it not only initializes the device pose but also corrects the accumulative errors at a certain frequency. However, the process of relocalization requires more than 60 ms on smartphones, which further makes it impossible to accomplish the pose estimation task along with the human keypoint detection task in real-time on mobile devices.

The art of edge computing supplies a promising idea, decoupling a complex task into two pipelines, the pipeline with intensive computation is offloaded to the edge, and the other with lightweight tasks runs on the mobile device. Inspired by this offloading strategy, as illustrated in Fig. 8.2, it is easy to design an edge-assisted basic solution for our human keypoint detection task and pose estimation task. Specifically, the pose estimation task is decoupled into short-term VO running on the mobile device and relocalization running on the edge. The human keypoint detection task is decoupled into human keypoint tracking on the mobile device and human keypoint detection on the edge. To better understand the challenge, we model the

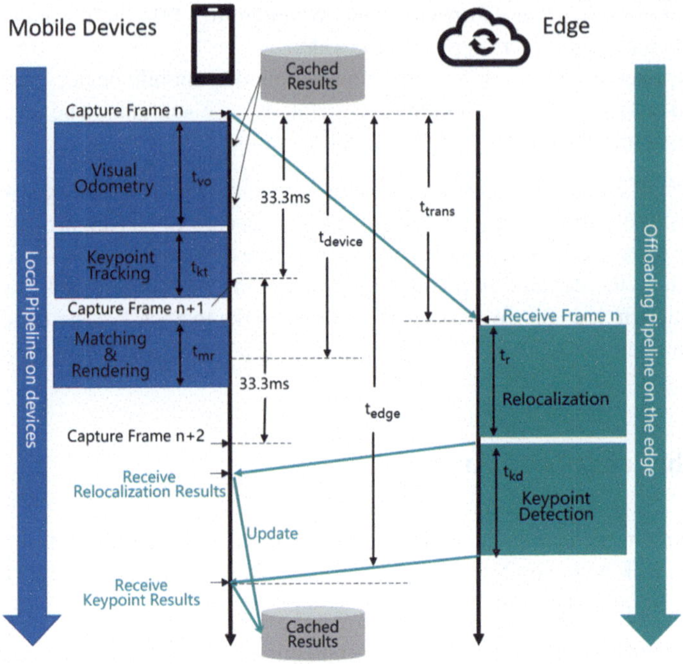

Fig. 8.2 Latency analysis of the basic solution

motion-to-photon latency on the mobile device of the basic solution as follows:

$$t_{device} = t_{vo} + t_{kt} + t_{mr}$$
$$t_{edge} = t_{trans} + t_{kd} + t_{rl}. \tag{8.1}$$

As shown in Fig. 8.2, the mobile AR devices (i.e., smartphone or AR HUD) are assumed to be connected to an edge server through a wireless connection (i.e., Wi-Fi or LTE). When capturing a new frame n, the local pipeline compares it with cached results on the former frames to update the device pose and human keypoints for matching and rendering processes. Thus, the latency of local pipeline on devices t_{device} mainly includes processing latency of visual odometry t_{vo}, human keypoint tracking t_{kt}, and matching and rendering t_{mr}. The latency involved in offloading mainly consists of three components: the time taken to transmit frame n to the edge server (t_trans), the time required to run human keypoint detection (t_kd), and the time for visual relocalization (t_re) on the edge server. After the edge finishing the human keypoint detection and relocalization tasks, the mobile device will receive the latest results transmitted back from the edge and update the cached results. As illustrated in Fig. 8.3a, running both VO and human keypoint tracking already consumes more than 40 ms on mobile devices ($t_{vo} + t_{kt} > 40$ ms), and there is no time left for matching their results and rendering on the screen, which obviously

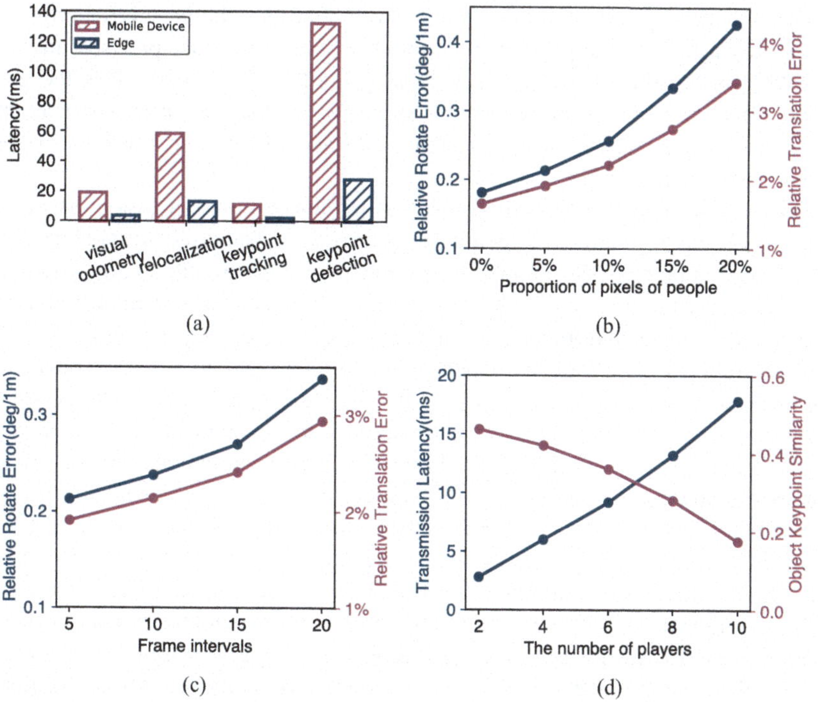

Fig. 8.3 Latency and accuracy analysis of the basic solution. (**a**) Latency of each function unit of the basic solution. (**b**) Impacts of environmental dynamics for visual odometry. (**c**) Accumulative error of on-device visual odometry. (**d**) Impact of multiple users on latency and accuracy

cannot satisfy the real-time constraint of the on-device pipeline ($t_{device} < 33.3$ ms). It is the overarching challenge to optimize the on-device pipeline and ensure the real-time constraint. It is worth noting that separately running single VO task or human keypoint tracking task on the mobile device can satisfy the real-time constraint (i.e., $t_{vo} + t_{mr} < 33.3$ ms and $t_{kt} + t_{mr} < 33.3$ ms), which indicates the extra latency produced by running dual tasks is the major factor. Fortunately, we observe the intrinsic similarity of above two tasks, that is, they both track image characteristic through consecutive video frames, which inspires us to merge similar computation to decrease the latency.

Accuracy Analysis Another critical challenge of the basic solution is the accuracy of the dual tasks, whose low accuracy will cause terrible interaction experience as for AR application. On the one hand, the pose estimation task suffers from accumulative error caused by the on-device VO which utilizes structure from motion (SfM) [16] to recover relative poses from consecutive frames. Specifically, environmental dynamics (mainly come from the moving users) will influence SfM to output relative pose with deviation, and the accumulated pose error can be eliminated only when receiving an absolute device pose from the relocalization module on the edge. Figure 8.3b quantitatively analyzes the pose error impacted by moving users. As the proportion of moving users grows up (pixels of moving users/all pixels), the pose error of VO rises significantly which makes the pose estimation task invalid. Besides, as shown in Fig. 8.3c, when the interval of receiving relocalization results from the edge increases, the accumulative error will be so large as to disable the pose estimation task.

On the other hand, the mobile device has no ability to detect new users before receiving the updated human keypoint detection results. Therefore, to avoid negative false detection in the basic solution, the mobile device has to offload every frame to the edge server because any fresh captured frame may contain new appeared users. Unfortunately, since multiple users offload key frames to the edge server through the wireless network, their highly frequent offloading will lead to network congestion and increase transmission latency(t_{trans}). It will extend t_{edge} and cause the accuracy of on-device human keypoint tracking dropping since when the device receives the result of human keypoint detection, the results have been out of date for several frames. For quantitative analysis, we have measured the transmission latency (t_{trans}) and human keypoint tracking accuracy as the number of users grows under the Wi-Fi 2.4 GHz and Wi-Fi 5 GHz networks. As illustrated in Fig. 8.3d, the basic solution performs sensitive to the multiple players scene, when the number of users increases to 10, the transmission latency reaches up to near 20 ms, and the human keypoint tracking accuracy (evaluated by Object Keypoint Similarity) drops down to 0.2, which will cause intolerable interaction experience for multi-user AR applications. However, according to statistics on over 2 hours of recorded video for multi-user AR applications, the number of frames those contain new appeared users who do not exist in several former frames accounts for only 6%. From a different perspective, it is a chance for us to reduce the offloading frequency and support more users if the user can foresee whether new users will appear and only offload these frames to the edge.

Matching Analysis The output of pose estimation is a 6-DOF expression containing a 3-DOF location and a 3-DOF rotation in the world coordinate, but the output of human keypoint detection is pixels annotated with labels located on a 2D image, which makes it difficult to connect these two locations without a united coordinate. Let us say in Fig. 8.4, if we want to match the device pose (T_2, R_2) of $user_2$ with his/her human body on $user_1$'s screen, first we need to transform it from the world coordinate to a 2D location (u_2, v_2) on $user_1$'s screen. And the process of transformation is described as follows:

$$\begin{bmatrix} u_2 \\ v_2 \\ 1 \end{bmatrix} = \begin{bmatrix} \frac{1}{dx} & 0 & u0 \\ 0 & \frac{1}{dy} & v0 \\ 0 & 0 & 1 \end{bmatrix} \begin{bmatrix} f & 0 & 0 \\ 0 & f & 0 \\ 0 & 0 & 1 \end{bmatrix} \begin{bmatrix} R_1 | T_1 \end{bmatrix} \begin{bmatrix} T_2 \\ 1 \end{bmatrix}. \tag{8.2}$$

dx, dy, v_0, u_0, f are known camera internal parameters of $user_1$ and (T_1, R_1) is the device pose of $user_1$.

After acquiring (u_2, v_2), the basic solution should match it with the corresponding human body. Since there are probably multiple users appearing on $user_1$'s

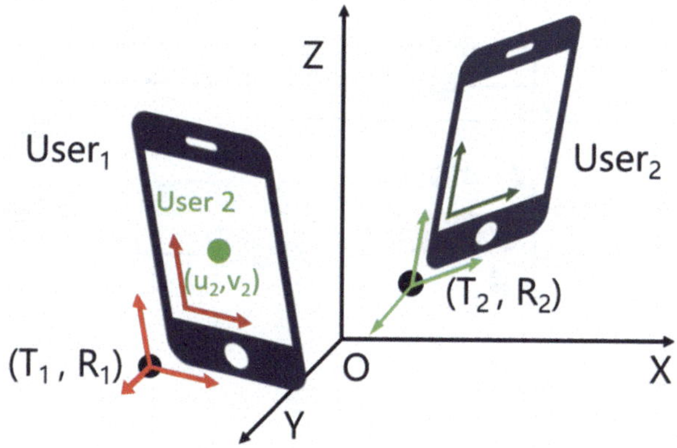

Fig. 8.4 3D–2D projection of device pose

screen, say $user_2$ and $user_3$, then we have to match (u_2, v_2) and (u_3, v_3) with two human bodies and endow the human bodies with their IDs. Considering two human bodies are close or overlapped on $user_1$'s screen, it becomes quite difficult to distinguish them. For example, the basic may mark the $user_2$ as $user_3$ on $user_1$'s screen which makes $user_1$ confused about their ID or interacts with the wrong user unconsciously.

8.3 System Architecture for EdgeSLAM-Based Multi-user AR

Figure 8.5 sketches the system architecture of DiTing in the multi-user AR scene where multiple users hold their smartphones to enjoy their real-time ID-aware visual interaction with other users appearing on their screens. On the top level, DiTing reforms the basic solution and proposes three key innovative designs to overcome all aforementioned challenges and enable real-time ID-aware interaction, which will be introduced in turns according to the workflow. First, when an AR user, say $user_3$, captures frame n, *Shared On-device Tracking* will leverage a hybrid optical flow tracking method to separately track the keypoints in moving human body and static background (environment) , which provides association information both for on-device VO and for human keypoint tracking. *Shared On-Device Tracking* merges the common computation of VO and human keypoint tracking, which saves the majority of time tracking feature points repeatedly. Then, *Tightly Coupled Dual Pipeline* improves and fuses two independent tasks (human keypoint detection and pose estimation) through a well-designed tightly coupled strategy which enables each task to benefit from mutual assistance. On the one hand, VO not only outputs the updated pose of $user_3$ but also helps foresee whether $user_1$ or $user_2$ will enter

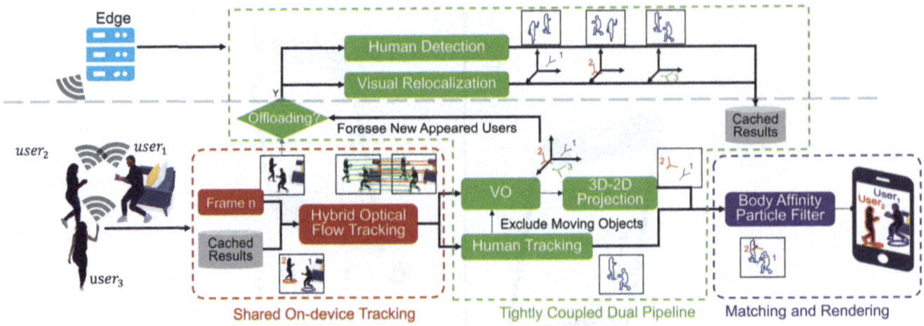

Fig. 8.5 System overview of DiTing

$user_3$'s screen according to their device poses, which decides whether to offload the next frame $n + 1$ to the edge for human keypoint detection. On the other hand, human keypoint tracking not only updates the human keypoint results on $user_3$'s screen, say $user_1$ or $user_2$, but also helps exclude the dynamic feature points in moving human bodies when calculating relative pose in VO module. Finally, *Body Affinity Particle Filter* matches device poses and human bodies, focusing on solving mismatching problems when multiple users, say $user_1$ and $user_2$, appear closely on $user_3$'s screen. In this way, DiTing can render keypoints on the human body according to his/her ID information and enable $user_1$ to visually interact with $user_2$ and $user_3$ on the screen.

8.4 Shared On-Device Tracking with EdgeSLAM

In this section, we explain the rationale for why we can merge two seemingly disparate on-device tasks, visual odometry and human keypoint tracking.

8.4.1 Device Pose Estimation

In general, the visual SLAM is widely used to achieve the pose estimation which can be divided into the indirect method and the direct method according to the different image information collected by the visual odometry [17]. For the sake of the combination with the human keypoint tracking, we choose to use the direct method visual SLAM (DSO [13]) to estimate the device pose which operates directly on the pixel intensity without any preprocessing. Specifically, the visual SLAM mainly consists of two key modules: front-end visual odometry and back-end relocalization. As shown in Fig. 8.5, we assign the front-end visual odometry (VO) to the mobile devices and offload the back-end relocalization to the edge server. The visual odometry module continuously tracks the device pose from consecutive

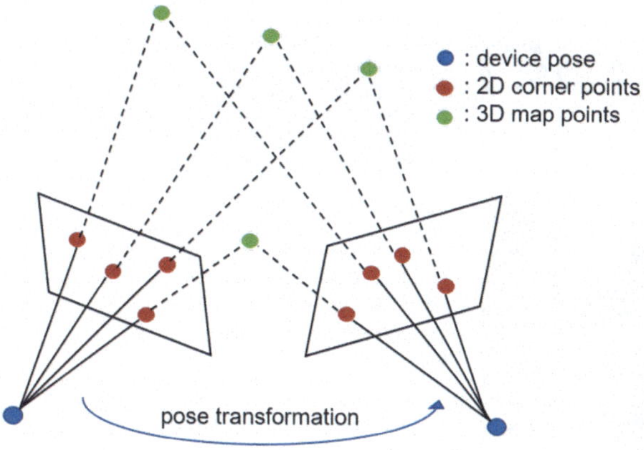

Fig. 8.6 Illustration of visual odometry

video frames. Specifically, when capturing a new video frame, we will extract some special pixels (corner points) and associate them with the corner points on the previous frame. Figure 8.6 shows an example of the association of extracted corner points between two frames. Donate the relative pose motion between the two frames as T, which can be easily computed by solving the so-called epipolar constraint equation [18]. Obviously, the key to estimate the relative pose is how to associate and match the corner points between the two frames.

8.4.2 Merge Two Tracking Tasks

In essence, the human keypoint tracking is to match the human keypoint of the two adjacent frames as the same as the key step of visual odometry. Therefore, we propose *Shared On-device Tracking* technique to merge the workflow of on-device VO and on-device human keypoint tracking by reusing the tracking results of feature points, which nearly halves the on-device computation time compared with the basic solution, thereby leaving enough time for the matching and rendering process to enable the real-time local pipeline on devices within 33.3 ms. On-device VO correlates the corner points of consecutive frames and adopts SfM (structure from motion) [16] to recover device poses from image sequences. Coincidentally, human keypoint tracking is designed to track the keypoints on the human body through consecutive frames annotated by the human keypoint detection results from the edge. As shown in Fig. 8.7, *Shared On-device Tracking* applies hybrid optical flow to track all feature points, including corner points (outside human body) and keypoints (inside human body). To save the computation of on-device VO and human keypoint tracking, *Shared On-device Tracking* utilizes optical flow to track all feature points in only one-time calculation, whose results are fed to the corresponding task.

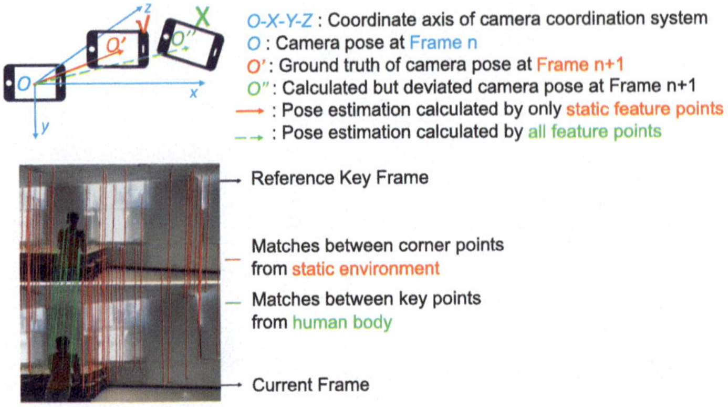

Fig. 8.7 Process of shared on-device tracking

8.5 Tightly Coupled Dual Pipeline Under EdgeSLAM

Tightly Coupled Dual Pipeline tightly fuses two pipelines, helping reduce the offloading frequency and supporting more users to join the interaction through mutual assistance. The accuracy of pose estimation and human keypoint tracking is crucial to matching and rendering processes; however, they both face their own predicaments. The accuracy of VO is affected by environment dynamics mainly caused by the moving users; hence, multi-user AR scene will cause serious pose drifts when calculating relative device pose. Human keypoint tracking fails to detect newly appearing users; therefore, it depends on frequently offloading video frames to the edge for keypoint detection. This approach significantly strains wireless bandwidth as the number of users grows.

To improve the accuracy of on-device VO, *Tightly Coupled Dual Pipeline* utilizes the human body information of the human keypoint detection task to exclude corner points on moving humans and promotes VO to use static corner points to calculate relative pose. To reduce the offloading frequency, *Tightly Coupled Dual Pipeline* utilizes the location information of the pose estimation pipeline to foresee the appearance of new users, that is to say, the user offloads frames to the edge only when new users are foreseen to enter his/her screen. As shown in Fig. 8.8, from $user_3$'s perspective, supposing we have obtained all four users' poses $\mathbf{P} = \{\mathbf{p}_1, \mathbf{p}_2, \mathbf{p}_3, \mathbf{p}_4\}$, we can project p_1, p_2, and p_4 (i.e., light-of-sight, LOS) to 2D coordinates on his/her screen by Eq. 8.2. To foresee new users appearing on $user_3$'s screen, we set a virtual fence around the LOS area called LOS fence and classify all the other users into three categories: the first like $user_2$ appears in the LOS area, which can be tracked by local human keypoint tracking; the second like $user_4$ is far from LOS area and outside LOS fence, and the last like $user_1$ is outside LOS area but inside (or partly inside) LOS fence. By the LOS fence design, *Tightly Coupled Dual Pipeline* utilizes pose estimation results to foresee the

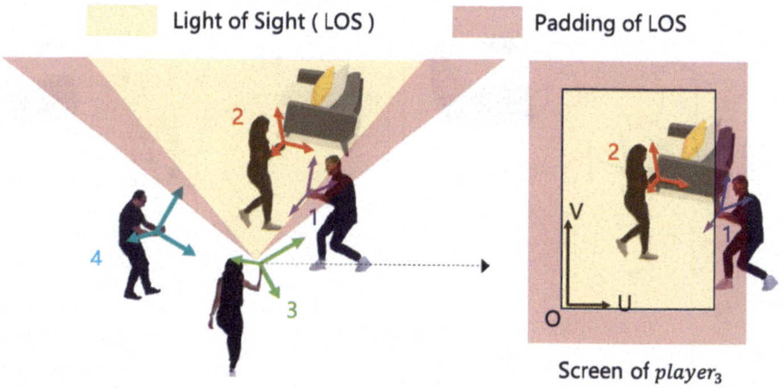

Fig. 8.8 Foresee new appeared users on the screen of $user_3$

new appeared users, in which case a user offloads the current frame to the edge for human keypoint detection, relieving the pressure on bandwidth and increasing the potential of supporting more users involved in AR interaction.

8.6 Body Affinity Particle Filter for Enhancing EdgeSLAM

Body Affinity Particle Filter can match the human bodies on the screen with the device poses and endow each human body with its corresponding user ID. During the matching process, it happens frequently that multiple users concurrently appear and stay closely on one's screen, which presents big challenges to precise matching and leads to confusing interaction. DiTing fixes this problem by using body affinity to measure the relationship between human bodies and device poses and then adopts an augmented particle filter based on body affinity to achieve precise matching between them.

Body Affinity As shown in Fig. 8.9a, body affinity describes the affinity between π particles $\mathbf{X} = \{X_i\}, 1 \leq i \leq \pi$, and transformed N user poses $\mathbf{P} = \{P_j = (u_j, v_j)\}, 1 \leq j \leq N$. Specifically, the body affinity $ba_{i,j}$ between particle X_i and the transformed pose P_j of $user_j$ is defined below:

$$ba_{i,j} = \frac{\sum_{k=1}^{\pi} \sqrt{(u_{pj} - u_k)^2 + (v_{pj} - v_k)^2}}{\pi \sqrt{(u_{pj} - u_i)^2 + (v_{pj} - v_i)^2}}. \tag{8.3}$$

Obviously, the particle X_i closer to the P_j will have higher $ba_{i,j}$, which means P_j has higher affinity with the human body containing particle X_i.

Particle Movement We use relative pose calculated by VO to design particle movement. As shown in Fig. 8.9b, the blue arrow represents the particle movement

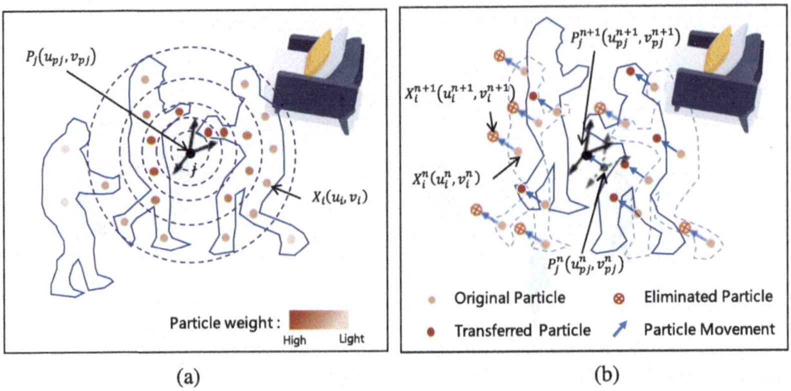

Fig. 8.9 Particle weight assignment and movement. (**a**) Particle weight assignment on frame n. (**b**) Particle movement on frame $n + 1$

between frame n and frame $n + 1$, which is calculated by the relative pose between the two frames and updates the location of particle X_i:

$$\begin{cases} u_i^{n+1} & = u_i^n + (\rho + \delta_i^{n+1}) \cos(\theta_i^{n+1} + \vartheta_i^{n+1}) \\ v_i^{n+1} & = v_i^n + (\rho + \delta_i^{n+1}) \sin(\theta_i^{n+1} + \vartheta_i^{n+1}) \\ (\rho \cos\theta, \rho \sin\theta) & = (u_{pj}^{n+1} - u_{pj}^n, v_{pj}^{n+1} - v_{pj}^n), \end{cases} \tag{8.4}$$

where P_{pj}^{n+1} and P_{pj}^n are calculated through Eq. 8.2 according to the poses at the time of frame n and frame $n + 1$, and δ_i^{n+1} is zero Gaussian noise to compensate the variations of relative pose estimation. Then, we check whether the particle after movement still stays in the original human body and then eliminate those moving outside. For each valid particle, we further update its weight.

Particle Weight Assignment Initially, all particles are equally weighted, i.e., $w_i = \frac{1}{\pi}, 1 \leq i \leq \pi$. After updating particle locations, particle weights are changed according to body affinity values (as shown in Fig. 8.9a). Specifically, the weight of ith particle is updated as

$$w_i = w_i * e^{\frac{ba_{i,j}^2}{2 * \sum^2}}, \tag{8.5}$$

where $\sum$ is the normalized parameter and set to 0.8 in our experiments.

Particle Resampling After updating the particle weights, we resample particles according to their updated weights, whose process is as follows:

Normalize the particle weights: $w_i = \frac{w_i}{\sum_{j=1}^{\pi}}$, where π is the total number of transferred particles in Fig. 8.9b.

Randomly generate a value $k, k \in [0, 1)$, and choose the corresponding index n subject to $\sum_{1 \le j < n} w_j \le k < \sum_{1 \le j < n+1} w_j$.
Create a new particle the same as X_n.
Repeat Steps 1 and 2 until the number of particle recovers to that before the movement.

Matching Decision Strategy The weights of distributed particles reflect the likelihood of the real location of the device pose on the screen. Selecting the particle with the highest weight and calculating a weighted average of all particles as the target location are two traditional ways. However, we find the former approach quickly converges but often leads to fluctuating matching, and the latter approach offers accurate matching but takes several frames to converge. Thus, we absorb two common approaches by first choosing the top-weighted 10% particles, then dividing them by the human body they belong to, and last matching the device pose with the human body containing maximum particles. In this way, we can achieve fast and accurate matching, endowing each human body with its ID in real-time.

Rationale Behind *Body Affinity Particle Filter* owes its high accurate and real-time matching to three aspects:

- Body affinity continuously describes the affinity relationship between body instances and device poses, which works in complex scenes, e.g., when the transferred device pose is outside any instance body or inside two overlapped instance bodies.
- DiTing uses the relative pose but not the absolute pose as the fundament of particle movement, which eliminates the effect of accumulative error of pose estimation.
- DiTing uses augmented decision strategy to match body instances and device poses, which avoids waiting latency and ensures stably accurate results.

8.7 System Implementation

The implementation of DiTing follows the system workflow in Fig. 8.5 and is developed by adopting two key tools *DSO* and *Mask R-CNN* [9]. Specifically, we adopt the server–client structure by arranging the VO and human keypoint tracking tasks on the client (mobile devices) and the relocalization and human keypoint detection tasks on the server (edge server).

Mobile Devices The implementation of DiTing follows the workflow in Fig. 8.5. We implement the local part of DiTing on four mobile devices (Nvidia Jetson TX2, Google Pixel 3, Galaxy S10, and iPhone 12). In the beginning, we continuously capture video frames from the camera on the mobile device using OpenCV API and feed it to *Shared On-device Tracking* module which is developed based on *DSO* [13]. We achieve hybrid optical flow tracking thread which extends the function implemented by `CoarseTracker` in *DSO*. It can not only track corner points for VO tracking but also complete human keypoint tracking meanwhile. Specifically, `findCornerPoints()` function extracts corner points for camera pose estimation. Then `filterPointsOnHuman()` function will remove the corner points lying on the human body for reducing the side effect causing by moving human. What is more, `goodFeaturePoints()` function will help us to filter some feature points very close together which can balance the processing latency and tracking accuracy. All feature points, including filtered corner points beyond human bodies and keypoints on human bodies, are tracked by `calcOpticalFlowPyrLK()` implemented by the classic LK optical flow algorithm. It is worth noting that considering the real-time requirements of optical flow algorithms on mobile devices, we have opted for the traditional LK optical flow algorithm instead of deep neural network-based optical flow methods. In *Tightly Coupled Dual Pipeline* module, the frame offloading decision function `isKeyFrame()` will be performed if there is/are new user(s) appeared or the visual odometry thread needs relocalization. Finally, the program enters *Matching and Rendering* module, the `bodyParticleFilter()` function will perform to match human body with the user's ID, and we can finish the rending eventually (as shown in Fig. 8.10a).

Edge Server The edge server is equipped with Lenovo IdeaPad-Y700 with i7-6700HQ CPU of 2.6 GHz main frequency and 16 G RAM, running the Ubuntu 16.0.4 operating system. The server's human keypoint detection program is developed on Mask R-CNN, and we apply our Mask R-CNN models with the ResNet-50 backbone and the network parameters are pretrained on COCO image dataset [19], and for Mask RCNN, the GPU we use is TITAN V with cuda-10.0.130 and cudnn-7.6.3. And the server's relocalization program is developed on edgeSLAM [15], and we utilize the relocalization module of edgeSLAM with some modifications. Figure 8.10b displays the interface through which the server maintains the global map and tracks the user's current pose.

8.8 Evaluation for EdgeSLAM-Based Multi-user AR

In this section, we first present the experiment setup (Sect. 8.8.1), followed by the overall performance of DiTing compared with SOTA systems in terms of overall accuracy and end-to-end latency (Sect. 8.8.2). We then conduct an ablation study to understand each functional module in DiTing. The main findings are our system is able to achieve both the high accuracy and the real-time requirement for two on-device tasks required by the multi-user AR scenario. Our system achieves 175% higher overall accuracy than the state-of-the-art baseline.

Fig. 8.10 Using case of DiTing: a first-person shooting multi-user augmented reality game. (**a**) Interface on the mobile AR device: the user wearing a virtual helmet. (**b**) Interface on the server: global map and user poses

(a)

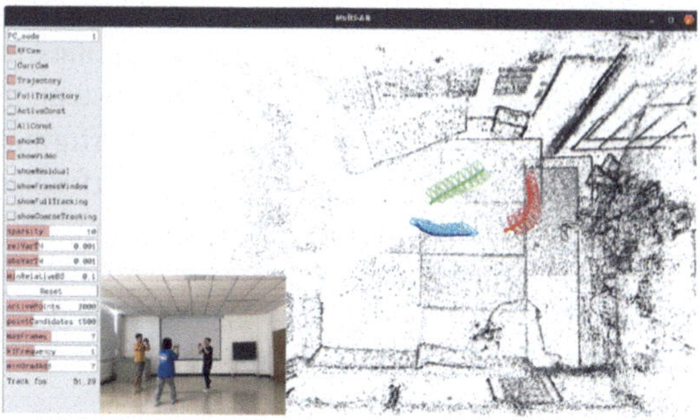

(b)

8.8.1 Experiment Setup

Datasets To extensively evaluate the overall accuracy of two tasks and the accuracy of every single task, we perform the experiences with different counterpart datasets.

- *Dataset for two tasks:* Due to the existing public datasets not covering both pose tracking and human keypoint tracking at the same time, we have made the dataset for the overall evaluation of two tasks. During the making of our dataset, we run the Mask R-CNN offline to obtain the human keypoint ground truth for every frame. And we run the ORB-SLAM2 offline (we eliminate the moving feature points on human body to guarantee the optical performance) to obtain the device pose information as the pose tracking ground truth. Our dataset covers multiple scenarios with the number of users ranging from 2 to 8.
- *Dataset for pose tracking:* We carefully choose five sequences from the odometry benchmark KITTI [20] dataset containing plentiful moving objects which is a well-known and challenging dataset for pose tracking.
- *Dataset for human keypoint tracking:* We as well choose five sequences from the PoseTrack18 dataset [21] which is a large-scale benchmark containing multiple persons for human keypoint tracking.

Baselines To extensively evaluate the performance of our design, we compare DiTing with **DSO** and **edgeSLAM**. The former one is a real-time SLAM system in mobile devices regarding the device's coordinate as the center of users in the video. The latter one is a state-of-the-art edge-assisted SLAM system with human body detection in recent work [15].

Evaluation Metrics According to different evaluated tasks with different datasets, we have set the corresponding evaluation metrics:

- *Pose tracking accuracy:* Relative Rotation Error (RRE) is an essential evaluation indicator in the device pose tracking task.
- *Human keypoint tracking accuracy:* Object Keypoint Similarity (OKS) is a common evaluation indicator in the human keypoint tracking task, to quantify the closeness of the tracked human keypoints with the ground truth keypoints.
- *Overall accuracy of two tasks:* In our system, the tracked human keypoints have their IDs. The OKS can quantify the performance of human keypoint tracking, and its ID correctness can reveal the performance of pose tracking. In order to evaluate our two tasks comprehensively, we propose a synthetic metric, **ID-OKS**, defined by Eq. 8.6. Specifically, as shown in Fig. 8.11, for every tracked keypoint, only when its ID is correct ($T_p = 1$), then we calculate its closeness with the ground truth keypoint; otherwise, the closeness should be zero ($T_p = 0$).

$$ID - OKS_P = T_p * \frac{\sum_i exp(-d_{pi}^2/2S_p^2\sigma_i^2)\delta(v_{pi} = 1)}{\sum_i \delta(v_{pi} = 1)}. \tag{8.6}$$

8.8.2 Overall Performance

We first compare DiTing with DSO, edgeSLAM, our solution with only on-device real-time optimization technique (SOT), and our solution without the matching optimization (SOT+TCDP), to evaluate the overall performance.

8.8.2.1 Overall Accuracy Comparison

Our system is able to achieve higher accuracy under both Wi-Fi 2.4 GHz and Wi-Fi 5 GHz network conditions. Figures 8.12a and b shows the CDF of ID-OKS for overall accuracy of two tasks. We draw two accuracy thresholds widely used in the computer version community: 0.5 as the loose threshold and 0.75 as the strict threshold; we mainly adopt the strict threshold in following deeper analysis. Specifically, under the Wi-Fi 2.4 GHz network, DiTing achieves 410% higher ID-OKS than DSO and 175% higher ID-OKS than edgeSLAM. Moreover, the results show that our three main techniques effectively increase the overall accuracy of our

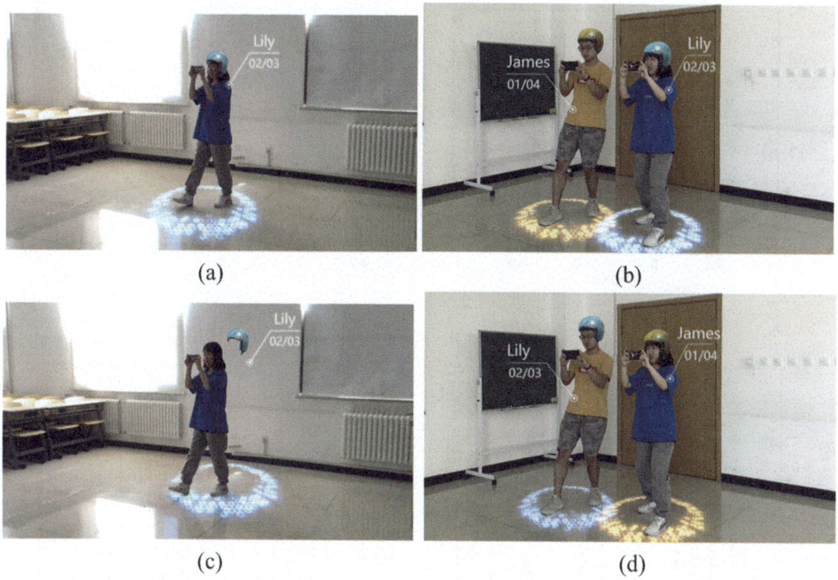

Fig. 8.11 IDs are matched right in (**a**,**b**) based on DiTing and wrong in (**c**,**d**) based on the baseline solution. (**a**) ID-OKS = 0.95. (**b**) ID-OKS = 0.95. (**c**) ID-OKS = 0.55. (**d**) ID-OKS = 0

system. The solution only with Shared On-device Tracking (SOT) still outperforms the edgeSLAM by 21.4%, which reveals our real-time tracking module in the mobile device is critical and is more effective than the tracking module of edgeSLAM. The solution further improves by 28.2% after integrated Tightly Coupled Dual Pipeline (TCDP). And there are equivalently significant improvements under the Wi-Fi 5 GHz network.

In addition, we further study the impact of different numbers of player on system performance. Results further show our system is less affected by the number of users increasing. As illustrated in Figs. 8.12c and d, with the number of players growing, the performance of all has a certain drop. With more players, there are less feature points in rigid contexts for help estimating the device pose, and at the same time it is more difficult to track so many moving feature points on human body. Fortunately, the ID-OKS of RIDAR still can reach 0.53 even with eight players under Wi-Fi 5GHz network, and it is relatively high in the human keypoint tracking task. And when the number of users grows from 4 to 8, the accuracy of edgeSLAM drops by 30.3% on average, but DiTing only drops by 19.7%, which shows the stronger capability to accommodate more users. We notice when the number of users reaches 8, the ID-OKS of DSO drops to only 0.08 because with more users, it is more likely to confuse and mismatch the device poses with the users in the video. We also find that when only with Shared On-device Tracking technique, its performance falls sharper with the increase of users than the system with Tightly Coupled Dual Pipeline technique, the reason behind is if without Tightly Coupled Dual Pipeline

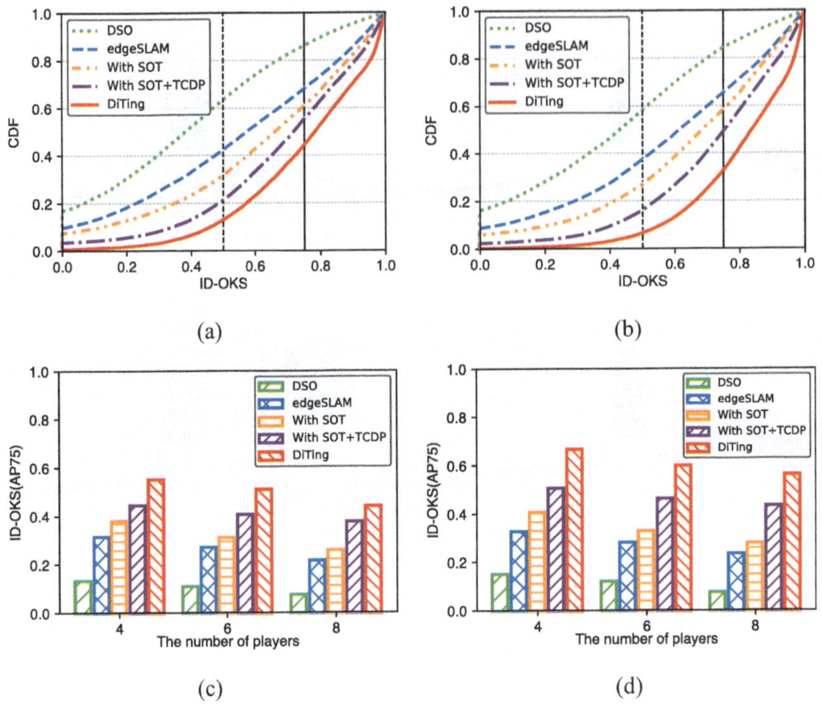

Fig. 8.12 Overall performance comparison under different network conditions. (**a**) Overall Accuracy—Wi-Fi 2.4GHz. (**b**) Overall Accuracy—Wi-Fi 5GHz. (**c**) Overall Accuracy—Wi-Fi 5GHz. (**d**) ID-OKS(AP75)—Wi-Fi 5GHz

technique, the mobile client should upload each frame once receiving the instance segmentation results from server due to lack of the information about other users' location. Therefore, the transmission latency will increase faster with more users which results in a rapid decrease in accuracy.

8.8.2.2 End-to-End Latency Comparison

We further examine the end-to-end latency of the mobile device including three key modules: on-device tracking, particle filter, and rendering. To validate our system is able to achieve an end-to-end latency within 33.3 ms inter-frame time at 30 fps to ensure a smooth VR game experience, we run our system under 2.4GHz Wi-Fi and calculate the average end-to-end latency for each frame in Fig. 8.13a. The black dashed line is the 33.3 ms deadline. The left bar is the latency of DiTing, and the right bar is the latency of the basic solution as described in Sect. 8.2. We can observe the device is able to finish all tasks including on-device tracking, particle filter, and rending within 33.3 ms under the major wireless connection conditions. What is

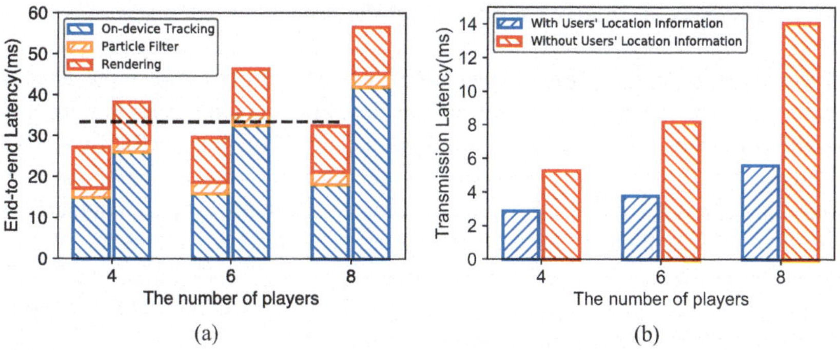

Fig. 8.13 Latency comparison with a growing number of players. (**a**) End-to-end latency. (**b**) Transmission latency

more, the average latency of our system only increases merely < 6 ms with the number of users increasing from 4 to 8, and by contrast, the average latency of the basic solution ranges from 38.4 to 66.7 which is far beyond the 33.3 ms deadline. The rapid increase of latency of the basic solution is mainly due to the time cost of on-device tracking, and it well evidences Shared On-device Tracking technique plays an important role in ensuring on-device real-time, especially when plenty of users are involved.

8.8.2.3 Transmission Latency Comparison

As we mentioned in Sect. 8.8.2.1, the long transmission latency will increase fast with the growing number of users. Then we measure the transmission latency with and without users' location information. As shown in Fig. 8.13b, the transmission latency will exceed 14 ms which takes up half the processing time if without users' location information. Furthermore, the average transmission latency reduces by 45.3% when four users are involved and greatly reduces by 60.3% when eight users are involved. The reason why the more users, the larger reduction proportion of transmission latency is that more users will bring more terrible network congestion which can be cured by less offloading frequency with users' location information.

8.8.3 Ablation Study

The end-to-end latency comparison in Sect. 8.8.2.2 reveals the effectiveness of our Shared On-device Tracking technique, and we then conduct a more exhaustive ablation study to understand the effectiveness of remaining techniques in DiTing.

8.8.3.1 Analysis of Tightly Coupled Dual Pipeline

Our Tightly Coupled Dual Pipeline technique can effectively improve the accuracy of both the pose tracking task and human keypoint tracking task. For the pose tracking task, Fig. 8.14a shows the tracking accumulative bias (RRE) when with or without TCDP technique. Thanks to the human body information, the tracking accumulative bias is within 0.25 degree for 1 meter length traces, which outperforms 24% the solution without TCDP technique. For the human keypoint tracking task, as seen in Fig. 8.14b, the average OKS increases from 0.38 to 0.49, improving by 29% with the technique because of less interval of receiving accurate human keypoint detection results from the edge.

8.8.3.2 Analysis of Body Affinity Particle Filter

Body Affinity Particle Filter technique is mainly designed to improve the ID matching precision, and we measure the user ID matching precision with and without leveraging this technique at the different number of users. As illustrated in Fig. 8.15a, when there are four users involved in the game, the matching precision of them is 84.2% and 79.1%, we first can find that there still exists some mismatching of users even with this technique, and we analyze the experimental results and explain that is because the mobile clients sometimes track other users failed when they are moving too fast. In addition, the matching precision of them only differs by 5.1%, which reveals it is relatively infrequent to mix up users' IDs when a small quantity of users is involved. However, when the number of users increases to 8, the matching precision will gain 14.1% which helps maintain the matching precision at a relatively high level, ensuring a good interaction experience for users.

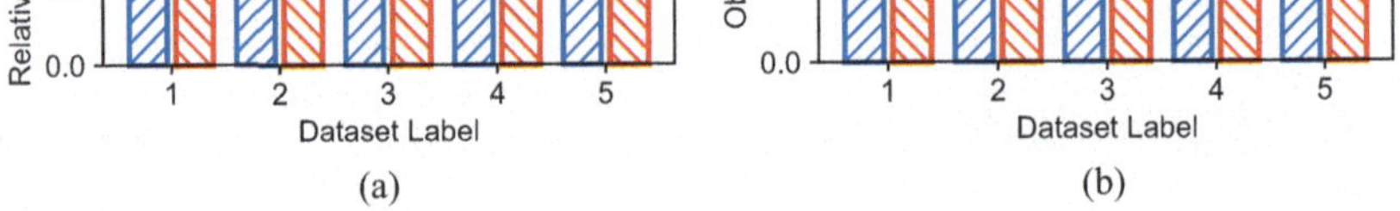

Fig. 8.14 The accuracy of pose estimation and keypoint tracking under public dataset w/o the TCDP technique. (**a**) Pose estimation accuracy. (**b**) Keypoint tracking accuracy

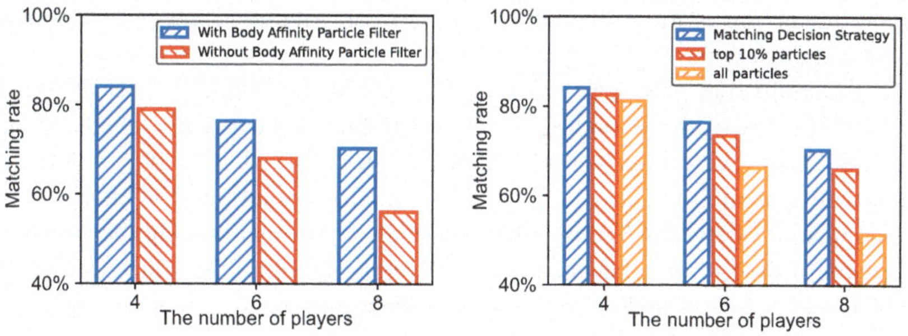

(a) (b)

Fig. 8.15 The ID matching precision w/o the BAPF technique or w/o the matching decision strategy. (**a**) ID matching precision w/o the BAPF technique. (**b**) ID matching precision with different matching strategy

8.8.4 Impact of Matching Decision Strategy

To demonstrate the effectiveness of designed matching decision strategy in Body Affinity Particle Filter technique, we evaluate the ID matching precision with leveraging the strategy or other two common strategies mentioned in Sect. 8.6. DiTing will select top 10% particles or all particles to calculate the target location if without our proposed matching decision strategy. As shown in Fig. 8.15b, the top 10% particles strategy will cause slight decline in ID matching precision because of without leveraging the human body information. Adopting the "all particles" strategy leads to a significant decline in ID matching precision. Moreover, this decline becomes more severe as the number of users increases. This issue likely arises because the matching process is time-consuming, resulting in delayed results. Our analysis indicates that the designed matching decision strategy is crucial for ensuring precise ID matching.

8.9 Related Work

8.9.1 Multi-user Augmented Reality

Augmented reality connects the physical world and virtual world, which offers the users an integrated interaction pattern, which means users can interact with their physical environment through the overlay of digital information [4, 5]. Several works focus on object detection for their AR system to generate virtual overlay [10, 22–24], while we adopt human keypoint detection which can acquire not only human bounding box regions like object detection but also the human joint information to support more complex AR application scenarios, and scilicet human keypoint detection is more powerful but more computational-intensive compared with object detection.

As for multi-user AR, when a user has gotten the results of detection (the pixel coordinates of the bounding boxes or the human keypoints), she/he should share the results with other users. CARS [25] discusses sharing object detection results between multiple users. SPAR [26] develops solutions for efficient data communications between AR users to reduce latency of maintaining the consistent placement of the virtual objects across users, while DiTing dexterously makes full use of the edge server as the center node to distribute results between users with little transmission overhead.

Different multi-user AR systems may work on different aspects of improving the user experience. CollabAR [27] aims to allow the users collaborate to recognize and annotate the virtual objects in their shared view. DeepFaceAR [28] combines CNNs alongside augmented reality to enable the users to recognize the faces of individuals and display some individual information. Marvel [29] is designed to offer a real-time annotation service, running on regular mobile devices while achieving low latency and low energy consumption. Brick [30] integrates several interaction patterns into a multi-user AR game at room scale, in which users collaborate to fill in a pattern of empty slots using digital "bricks" scattered about the room. While DiTing focuses on the user interaction experience by offering accurate matches between the virtual overlay on the user and his/her user identity.

8.9.2 Multi-device SLAM

For the multi-user AR system, pose estimation is another critical component in addition to object detection or human keypoint detection, which is usually performed by SLAM. Some SLAM systems only employ camera frames to estimate the device pose [12, 13, 15, 31], while other SLAM works additionally incorporate some odometry sensors, such as IMU, 2D laser scanner, or 3D LiDARs [32, 33]. As for DiTing, we aim at the multi-user mobile scenario; therefore, we adopt pure visual SLAM (DSO) without the help of any odometry sensor to eliminate any possible limitations of the users' device. Several multi-device SLAM works mainly focus on coordinate system alignment, which aims at developing efficient device pose communication framework. CCM-SLAM [34] transmits SLAM information whenever it is updated. SwarmMap [35] proposes a lightweight system service to reduce the transmission data volume. MISRA [36] utilizes a new VO network for pose estimation on mobile devices to ensure efficient pose sharing. While DiTing takes advantage of the guidance of the human keypoint detection which greatly reduces the frequency of unnecessary device pose transmission.

8.9.3 *Edge Assisted Real-Time Video Analysis*

Real-time video analysis plays a fundamental role in the wide spectrum of mobile smart applications, such as mobile AR, visual SLAM, and self-driving. Glimpse [22] is a continuous, real-time object recognition system for camera-equipped mobile devices, which uses a distributed architecture with the object recognition pipeline running at the server and an object tracking pipeline running on the devices. On that base, Liu et al. [10] further improve the end-to-end latency and enable high accuracy object detection for commodity AR systems running at 60fps. DeepDecision [24] puts forward an efficient "helper" on deciding whether to execute the simple model locally or the complex model remotely and makes better use of bandwidth by the adaption of video configuration. AdaScale [37] shows that rescaling the image to a lower resolution will sometimes produce better accuracy, which can accelerate the video object detection. On the other hand, Emmons et al. [38] open up the black box of neural networks, which joint inference between the cameras and the cloud, and continuous online learning for large deployments of cameras. DADS [39] allows adaptive partitioned DNN processed at both the edge and the cloud under network dynamics while limiting the data transmission. Recently, JCAB [40] proposes an efficient online algorithm that jointly optimizes configuration adaption and bandwidth allocation for multiple video streams connected to the same edge node. Edge-SLAM [41] and edgeSLAM [15] enable SLAM system for mobile devices, achieving both accuracy and real-time simultaneously with the help of edge computation. However, DiTing optimizes both pose estimation task and human keypoint detection task in multi-user AR applications, enabling the real-time inter-user interaction through a well-designed tightly coupled pipeline.

8.10 Conclusion

In this chapter, we design DiTing deeply integrating human keypoint detection and pose estimation, which is the first multi-user AR system supporting real-time ID-aware visual interaction. We prototype DiTing on four different commodity devices and develop a multi-user AR game as case studying, and the results show that DiTing runs at 30 fps, increases the accuracy of pose estimation and human keypoint detection by 24% and 29%, and achieves 175% higher overall accuracy than the state-of-the-art baseline.

References

1. L. Zhang, A. Sun, R. Shea, J. Liu, M. Zhang, Rendering multi-party mobile augmented reality from edge, in *Proceedings of the 29th ACM Work shop on Network and Operating Systems Support for Digital Audio and Video*, 2019, pp. 67–72

2. A. Ortiz, C. Vitery, C. González, H. Tobar-Muñoz, Evaluation of an augmented reality multiplayer learning game, in *Joint International Conference on Serious Games* (Springer, 2018), pp. 91–100
3. S. Byvaltsev, Review of the features of augmented reality application in the training of operators and maintenance staff, in *IOP Conference Series: Materials Science and Engineering*, vol. 966 (IOP Publishing, 2020), p. 012 121
4. J. Paavilainen, H. Korhonen, K. Alha, J. Stenros, E. Koskinen, F. Mayra, The pokémon go experience: A location-based augmented reality mobile game goes mainstream, in *Proceedings of the 2017 CHI Conference on Human Factors in Computing Systems*, 2017, pp. 2493–2498
5. G. Ekaputra, C. Lim, K.I. Eng, Minecraft: A game as an education and scientific learning tool, *ISICO 2013*, vol. 2013, 2013
6. S.y.D. Hu, N. Wang, Multiplayer augmented reality: The future is social, presented by niantic, ser. SIGGRAPH '18, (Association for Computing Machinery, Vancouver, BC, Canada, 2018). ISBN: 9781450358217. [Online]. Available: https://doi.org/10.1145/3226552.3226585
7. J. Li, E.D. Van der Spek, X. Yu, J. Hu, L. Feijs, Exploring an augmented reality social learning game for elementary school students, in *Proceedings of the Interaction Design and Children Conference*, 2020, pp. 508–518
8. V. Kuts, T. Otto, Y. Bondarenko, F. Yu, Digital twin: Collaborative virtual reality environment for multi-purpose industrial applications, in *ASME International Mechanical Engineering Congress and Exposition*, vol. 84492 (American Society of Mechanical Engineers, 2020). V02BT02A010
9. chrispolo, *Keypoints-of-humanpose-with-mask-r-cnn* (2019). https://github.com/chrispolo/keypoints-of-humanpose-with-Mask-R-CNN
10. L. Liu, H. Li, M. Gruteser, Edge assisted real-time object detection for mobile augmented reality, in *Proceedings of the ACM MobiCom* (2019)
11. C. Forster, M. Pizzoli, D. Scaramuzza, SVO: Fast semi-direct monocular visual odometry, in *2014 IEEE International Conference on Robotics and Automation (ICRA)* (IEEE, 2014), pp. 15–22
12. R. Mur-Artal, J.M.M. Montiel, J.D. Tardos, ORB-SLAM: A versatile and accurate monocular slam system. IEEE Trans. Robot. **31**(5), 1147–1163 (2015)
13. J. Engel, V. Koltun, D. Cremers, *Direct sparse odometry* (2016). arXiv:1607.02565 [cs.CV]
14. Y. Huang, H. Zhao, X. Qiao, J. Tang, L. Liu, Towards video streaming analysis and sharing for multi-device interaction with lightweight DNNs, in *IEEE INFOCOM 2021 - IEEE Conference on Computer Communications* (2021), pp. 1–10. https://doi.org/10.1109/INFOCOM42981.2021.9488846
15. J. Xu, H. Cao, D. Li, et al., Edge assisted mobile semantic visual slam, in *IEEE INFOCOM 2020-IEEE Conference on Computer Communications* (IEEE, 2020), pp. 1828–1837
16. J.L. Schonberger, J.-M. Frahm, Structure-from-motion revisited, in *Proceedings of the IEEE Conference on Computer Vision and Pattern Recognition* (2016), pp. 4104–4113
17. W. Chen, G. Shang, A. Ji, et al., An overview on visual slam: From tradition to semantic. Remote Sens. **14**(13), 3010 (2022)
18. G. Xu, Z. Zhang, *Epipolar Geometry in Stereo, Motion and Object Recognition: A Unified Approach*, vol. 6 (Springer Science & Business Media, 1996)
19. T.-Y. Lin, M. Maire, S. Belongie, et al., *Microsoft Coco: Common Objects in Context* (2015). arXiv:1405.0312 [cs.CV]
20. A. Geiger, P. Lenz, R. Urtasun, Are we ready for autonomous driving? The KITTI vision benchmark suite, in *Proceedings of the IEEE CVPR* (2012)
21. M. Andriluka, U. Iqbal, E. Insafutdinov, et al., PoseTrack: A benchmark for human pose estimation and tracking, in *Proceedings of the IEEE Conference on Computer Vision and Pattern Recognition* (2018), pp. 5167–5176
22. T.Y.-H. Chen, L. Ravindranath, S. Deng, P. Bahl, H. Balakrishnan, Glimpse: Continuous, real-time object recognition on mobile devices, in *Proceedings of the ACM Sensys* (2015)

23. Q. Liu, T. Han, Dare: Dynamic adaptive mobile augmented reality with edge computing, in *2018 IEEE 26th International Conference on Network Protocols (ICNP)* (IEEE, 2018), pp. 1–11
24. X. Ran, H. Chen, X. Zhu, Z. Liu, J. Chen, DeepDecision: A mobile deep learning framework for edge video analytics, in *Proceedings of the IEEE INFOCOM* (2018)
25. W. Zhang, B. Han, P. Hui, V. Gopalakrishnan, E. Zavesky, F. Qian, CARS: Collaborative augmented reality for socialization, in *Proceedings of the 19th International Workshop on Mobile Computing Systems & Applications* (2018), pp. 25–30
26. X. Ran, C. Slocum, Y.-Z. Tsai, K. Apicharttrisorn, M. Gorlatova, J. Chen, Multi-user augmented reality with communication efficient and spatially consistent virtual objects, in *Proceedings of the 16th International Conference on emerging Networking EXperiments and Technologies* (2020), pp. 386–398
27. J. Stojkovic, Z. Liu, G. Lan, C. Joe-Wong, M. Gorlatova, Edge-assisted collaborative image recognition for augmented reality: Demo abstract, in *Proceedings of the 17th Conference on Embedded Networked Sensor Systems* (2019), pp. 394–395
28. A. Golnari, H. Khosravi, S. Sanei, DeepFaceAR: Deep face recognition and displaying personal information via augmented reality, in *2020 International Conference on Machine Vision and Image Processing (MVIP)* (IEEE, 2020), pp. 1–7
29. K. Chen, T. Li, H.-S. Kim, D.E. Culler, R.H. Katz, Marvel: Enabling mobile augmented reality with low energy and low latency, in *Proceedings of the 16th ACM Conference on Embedded Networked Sensor Systems* (2018), pp. 292–304
30. P. Bhattacharyya, Y. Jo, K. Jadhav, R. Nath, J. Hammer, Brick: A synchronous multiplayer augmented reality game for mobile phones, in *Extended Abstracts of the 2019 CHI Conference on Human Factors in Computing Systems* (2019), pp. 1–4
31. J.J. Tarrio, S. Pedre, Real-time edge-based visual odometry for a monocular camera, in *Proceedings of the IEEE International Conference on Computer Vision* (2015), pp. 702–710
32. R. Dubé, A. Gawel, H. Sommer, J. Nieto, R. Siegwart, C. Cadena, An online multi-robot slam system for 3D Lidars, in *2017 IEEE/RSJ International Conference on Intelligent Robots and Systems (IROS)* (IEEE, 2017), pp. 1004–1011
33. R. Chellali, et al., A distributed multi robot slam system for environment learning, in *2013 IEEE Workshop on Robotic Intelligence in Informationally Structured Space (RiiSS)* (IEEE, 2013), pp. 82–88
34. P. Schmuck, M. Chli, Multi-UAV collaborative monocular slam, in *Proceedings of the IEEE ICRA* (2017)
35. J. Xu, H. Cao, Z. Yang, et al., SwarmMap: Scaling up real-time collaborative visual slam at the edge, in *Proceedings of the USENIX NSDI* (2022), pp. 977–993
36. Y. Huang, H. Zhao, X. Qiao, J. Tang, L. Liu, Towards video streaming analysis and sharing for multi-device interaction with lightweight DNNs, in *IEEE INFOCOM 2021-IEEE Conference on Computer Communications* (IEEE, 2021), pp. 1–10
37. T.-W. Chin, R. Ding, D. Marculescu, AdaScale: Towards real-time video object detection using adaptive scaling (2019). Preprint. arXiv:1902.02910
38. J. Emmons, S. Fouladi, G. Ananthanarayanan, S. Venkataraman, S. Savarese, K. Winstein, Cracking open the DNN black-box: Video analytics with DNNs across the camera-cloud boundary, in *Proceedings of the 2019 Workshop on Hot Topics in Video Analytics and Intelligent Edges* (2019), pp. 27–32
39. C. Hu, W. Bao, D. Wang, F. Liu, Dynamic adaptive DNN surgery for inference acceleration on the edge, in *IEEE INFOCOM 2019-IEEE Conference on Computer Communications* (IEEE, 2019), pp. 1423–1431
40. C. Wang, S. Zhang, Y. Chen, Z. Qian, J. Wu, M. Xiao, Joint configuration adaptation and bandwidth allocation for edge-based real-time video analytics, in *IEEE INFOCOM 2020-IEEE Conference on Computer Communications* (IEEE, 2020), pp. 257–266
41. A.J. Ben Ali, Z.S. Hashemifar, K. Dantu, Edge-slam: Edge-assisted visual simultaneous localization and mapping, in *Proceedings of the ACM MobiSys* (2020)

Part IV
Conclusion

Part IV offers a conclusion and presents a contemplative look towards the horizon of edge-assisted visual SLAM. We synthesize the insights gleaned from previous discussions and outline the path forward, underscoring the emerging trends, potential applications, and the next frontiers awaiting researchers and practitioners in the field.

Chapter 9
Research Summary and Open Issues

Abstract In this chapter, we summarize this book and, more importantly, present a vision for future research directions. After almost 20 years of development in academia, visual SLAM systems have gained more attention from the industry, making their way toward robots, drones, and our daily lives. However, several challenges still need to be overcome before edge-assisted mobile visual SLAM is put into ubiquitous use.

9.1 Summary

In this book, we provide a thorough exploration of the synergy between visual SLAM and edge computing, tracing the trajectory from fundamental concepts to state-of-the-art applications. The monograph highlights the evolution of visual SLAM, initially burdened by computational intensity, and how edge computing emerges as a pivotal technology, enhancing real-time processing capabilities and enabling new frontiers for mobile devices.

The initial chapters lay the groundwork, detailing the intricacies of visual SLAM and its vital components. The discussion extends to the computational challenges on mobile platforms, setting the stage for the introduction of edge computing as a transformative solution.

Advancements in system architecture are then dissected, illustrating how edge computing dovetails with mobile visual systems. Innovations in scalability are presented, particularly for multi-agent systems, with SwarmMap serving as a hallmark development, showcasing its application in complex environments like oil-field explorations.

Subsequent sections delve into practical applications, from autonomous navigation and large-scale mapping to environment understanding. The narrative culminates in evolving multiuser augmented reality (AR) experiences, spotlighting shared on-device tracking with collaborative visual SLAM.

The insights presented in this volume chart the cutting-edge progress in edge-assisted visual SLAM, underscoring the transformative impact of integrating edge computing with mobile systems. Although strides have been made, a discernible

J. Xu et al., *Edge Assisted Mobile Visual SLAM*,
https://doi.org/10.1007/978-981-97-3573-0_9

divide persists between academic explorations and deployable real-world applications. Bridging this chasm demands concerted efforts from the global research community. It is the aspiration of this work to galvanize increased interest and collaborative endeavors from academic circles and industry sectors alike, propelling edge-assisted visual SLAM into a future rich with practical implementation and innovation.

9.2 Open Issues

Both edge computing and visual SLAM systems are evolving rapidly and continue to develop from diverse views. We summarize some of the interesting and attractive ones from the perspectives of *architecture*, *algorithm*, and *implementation*.

9.2.1 On the Architecture Front

The prevalent assumption in existing edge-assisted visual SLAM systems is the computational insufficiency and resource constraints of mobile devices. The architecture is thus designed primarily to offload tasks to the edge, allowing it to supplement the mobile device's capabilities. However, with the continuous advancements in chip design techniques and the integration of embedded chips (e.g., Xilinx Zynq), GPUs, and NPUs on mobile platforms, there is a substantial augmentation in mobile computational power. In this evolving landscape, a pivotal open issue and area for future work lies in upgrading system architecture. The goal is to transition from a paradigm where the edge merely aids mobile devices to one where there is true *mobile-edge collaboration*. Research could now pivot toward leveraging the increased computational prowess of mobile devices, orchestrating a more balanced workload distribution, and exploiting the nuanced capabilities of both the edge and mobile processors. This evolution promises a more robust, adaptive, and high-performing visual SLAM system, fit for the next generation of smart mobility and autonomous navigation technologies.

9.2.2 On the Algorithm Front

Two significant upgrades beckon our attention. The first is the embracing of deep learning technologies. The landscape of SLAM is gradually shifting from traditional optimization and filtering frameworks to deep learning-based methodologies, including Convolutional Neural Networks (CNNs) and Neural Radiance Fields (NeRFs). With mobile devices achieving finer granularity in scene modeling and a deeper understanding of environments, crafting effective algorithms that adapt edge-assisted visual SLAM to these latest developments is a critical research direction.

Second, in multi-agent collaboration, systems like SwarmMap currently only provide a foundational platform for map transmission and management in multi-agent SLAM. While SLAM map optimization remains a hot topic in robotics, integrating advanced technologies—such as map compression, fusion, and semantic recognition—into SwarmMap for enhanced performance is an ongoing effort. Moreover, efficient sharing of map data, not only between mobile and edge but also among different agents, could offer substantial benefits to upper-layer applications. These algorithmic enhancements are vital in propelling edge-assisted visual SLAM into a future where accuracy, efficiency, and collaborative intelligence are seamlessly intertwined.

9.2.3 On the Implementation Front

Existing edge-assisted SLAM and, indeed, most edge computing systems require specialized design and implementation tailored to specific software and hardware configurations. Any change in system setup, such as switching algorithms from VINS to ORB-SLAM or transitioning mobile hardware from smartphones to drones, necessitates significant manual effort for system reimplementation, revealing a lack of system flexibility and universality. Just as virtualization technologies (like Docker, Kubernetes, etc.) form the cornerstone of the cloud computing era, contemplating the construction of an effective virtualization platform tailored for edge computing emerges as a compelling direction. Such a platform is critical for accelerating the agile deployment of edge computing across a broader spectrum of devices, thereby enhancing system adaptability and paving the way for a more universally applicable edge computing paradigm.